保护海洋濒危动物丛书

龟去来兮

——走进海龟的世界

Marine Turtles: There and Back Again

Enter the Marine Turtles' World

王静　范敏　张宇　刘敏　主编

Editors: Jing Wang, Min Fan, Yu Zhang, Min Liu

中国海洋大学出版社

·青岛·

图书在版编目 (CIP) 数据

龟去来兮：走进海龟的世界 / 王静等主编 . —青岛：中国海洋大学出版社，2022.5
（保护海洋濒危动物丛书 / 杨海萍主编）
ISBN 978-7-5670-3164-7

Ⅰ. ①龟… Ⅱ. ①王… Ⅲ. ①海龟－濒危动物－动物保护－青少年读物 Ⅳ. ① Q959.6-49

中国版本图书馆 CIP 数据核字 (2022) 第 084878 号

出版发行 中国海洋大学出版社
社　　址 青岛市香港东路23号　　邮政编码 266071
网　　址 http://pub.ouc.edu.cn
出 版 人 杨立敏
责任编辑 邹伟真　　电　　话 0532-85902533
电子信箱 zwz_qingdao@sina.com
印　　制 青岛海蓝印刷有限责任公司
版　　次 2022 年 5 月第 1 版
印　　次 2022 年 5 月第 1 次印刷
成品尺寸 195 mm × 235 mm
印　　张 9.5
字　　数 90 千
印　　数 1 ~ 4000
定　　价 68.00 元
订购电话 0532-82032573（传真）

编 委 会

EDITORIAL BOARD

序言

海龟是一类古老的生物，从恐龙鼎盛时代一直存续至今。世界上现存有七种海龟，其中绿海龟、玳瑁、红海龟、太平洋丽龟和棱皮龟在我国海域有分布。在我国的传统文化中，海龟是吉祥、长寿的象征，在沿海地区被视为庇护平安的神灵。海龟拥有独特的生活史和习性，是海洋生态系统中重要的旗舰物种和指示物种，受到国际社会的高度关注。目前，世界现存的七种海龟均已被列入《濒危野生动植物种国际贸易公约》（CITES）附录Ⅰ。除了仅在澳大利亚分布的平背龟外，其他六种海龟也被列入《保护迁徙野生动物物种公约》（CMS）附录Ⅰ。在我国海域分布的海龟均在2021年初被升级为国家一级保护野生动物。保护海龟对于维护海洋生物多样性和生态系统平衡、促进自然与社会和谐发展具有特殊意义。

由于人类活动的影响和海洋生态环境的破坏，全球海龟承受巨大生存压力。我国海域内的海龟也存在栖息地衰退、种群数量下降等问题，物种延续面临挑战。但幸运的是，在习近平生态文明思想引领下，作为海洋生态系统旗舰物种的海龟受到全社会的广泛关注，一系列保护工作全面开展。2018年，原农业部牵头成立了中国海龟保护联盟，旨在通过搭建一个多方参与的平台，集合各方的共同努力推进海龟的保护工作，并在2019年发布了《海龟保护行动计划（2019—2033）》，为全国的海龟保护工作提供了指南。通过各方力量的不懈努力，海龟栖息地保护、人工繁育、规范救护和放归、公众参与等工作取得积极进展，特别是相关职能部门联合打击海龟非法捕捞和非法交易，海龟的生存状况得到了很大改善。

海洋是我们共同的家园，保护海洋、保护海龟需要你我共同的努力。作为曾经参与水生野生动物保护工作的一员，我很愿意为大家推荐《龟去来兮》这本书。该书不仅系统性地介绍了海龟的生态学、生命史等知识，还为我们展示了海龟保护工作者们在响应《海龟保护行动计划》和保护海龟过程中的各种努力，是一本集科学性、趣味性于一体的好书。希望该书能够让更多的朋友了解海龟和它们生活的环境，增加对海龟的认识，喜欢海龟，参与到海龟和海洋的保护工作中来。希望通过我们共同的努力，让海龟这类可爱吉祥的海洋精灵能够自由自在地生活在海洋家园，让这些古老且神秘的生命在地球上永续繁衍。

韩旭

2021年9月于北京

PREFACE

Marine turtles are a group of ancient creatures that have survived since the heyday of the dinosaurs. There are seven species in the world. Among these, five species occur in Chinese waters, including the green, the hawksbill, the loggerhead, the olive ridley and the leatherback. In traditional Chinese culture, marine turtles are symbols of auspiciousness and longevity and are regarded as gods of protection and peace by people who live in coastal areas. Marine turtles are important flagship and indicator species in marine ecosystems due to their unique life history and have received great attention internationally. All seven marine turtle species are listed in the Convention on International Trade in Endangered Species of Wild Fauna and Flora (CITES) Appendix I. Except for flatback turtles, which are only found in Australian waters, six species are also listed in the Convention on the Conservation of Migratory Species of Wild Animals (CMS) Appendix I. The five species that can be found in Chinese waters were upgraded to category I protected wild animals in early 2021. Marine turtle conservation is especially significant for maintaining marine biodiversity and ecosystem balance, and for promoting the harmonious development of nature and society.

Due to the impact of human activities and the destruction of the marine ecological environment, marine turtles around the world are facing tremendous challenges for survival. Marine turtles in China are also facing problems such as habitat loss and population decline. Fortunately, under the guidance of Xi Jinping's thoughts on ecological civilization, marine turtles, as a flagship species in marine ecosystems, have received extensive attention from relevant government departments and societies, and a series of conservation actions have been carried out comprehensively. The national fisheries administration took the lead in establishing the China Sea Turtle Conservation Alliance in 2018, which aims to build a multi participatory platform to promote marine turtle conservation, and released the "Sea Turtle Conservation Action Plan (2019－2033)" in 2019, which provides guidelines for the national conservation of marine turtles. Through the unremitting efforts of various parties, significant progress has been made in marine turtle habitat protection, captivity breeding, rescue and release protocol, and public engagement. In particular, relevant authority departments have joined together to combat illegal harvest and trade for marine turtles, hence their survival conditions have been greatly improved.

The ocean is our shared homeland, and to protect the ocean and marine turtles requires the joint efforts of you and me. As a member who has participated in aquatic wildlife conservation, I highly recommend the book *Marine Turtles: There and Back Again* to you. This book, combining science and pleasure, not only systematically introduces marine turtles, but also shows various efforts of various parties working on marine turtle conservation in response to the "Sea Turtle Conservation Action Plan". I hope that this book will enable more friends to understand marine turtles and their living environment and encourage people to participate in the protection of marine turtles and the ocean. Hopefully, marine turtles, such lovely and auspicious creatures, and ancient and magical lives, can live happily and thrive in their ocean homeland under our efforts.

Jun Xu

September, 2021 in Beijing

目 录

CONTENTS

第一章
Chapter 1

海龟知多少
About Marine Turtles

"年纪是我的闹钟。"——海明威《老人与海》

"Age is my alarm clock." (*The Old Man and the Sea* by E.M. Hemingway)

海洋里，有这样一类生物，古老、顽强且珍贵，陪伴着地球几经沧桑巨变，至今繁衍不息，它们就是被誉为地球“活化石”的——海龟。

Marine turtles, one of the “living fossils” found in the ocean, are an ancient, indomitable and precious group of species. They have thrived on this planet for hundreds of millions of years, witnessing the changes of the Earth over time.

第一章 Chapter 1

海龟知多少 About Marine Turtles

海龟在 1.5－2 亿年前就出现在了地球上，从恐龙鼎盛时代一直存续至今，是现存体型最大的海洋爬行动物之一。

海龟经历了从海洋到陆地又从陆地返回海洋的进化历程，拥有独特的习性和生活史，是海洋生态系统中重要的旗舰物种。

Marine turtles appeared over 150–200 million years ago, having survived from the heyday of the dinosaurs, and are one of the largest marine reptiles in the ocean today.

Marine turtles have undergone an evolution of moving from the ocean to the land and back to the ocean. They have unique behaviors and life histories and they are a flagship species of the marine ecosystem.

穿越亿万年与我们相遇 ▶
Through millions of years to meet us

第一章 海龟知多少

Chapter 1 About Marine Turtles

海龟广泛分布于太平洋、大西洋和印度洋。除了成年雌性海龟会在繁殖季节爬上沙滩挖坑产卵外，海龟的一生几乎都在海洋中度过。虽然可以下潜到水下几百米深的地方，但是海龟大部分时间还是在上层水域活动。不过，它们会在浅水的礁石或珊瑚丛中落脚休息和睡觉。

Marine turtles are distributed widely throughout the Pacific, Atlantic, and Indian Oceans. They spend their lives in the sea except for breeding females, who crawl onto the beach to lay eggs during the breeding season. Marine turtles can dive hundreds of meters; however, most of the time they prefer to hang around in shallow water, such as around rocky or coral reefs.

© 何云昊 Yunhao He

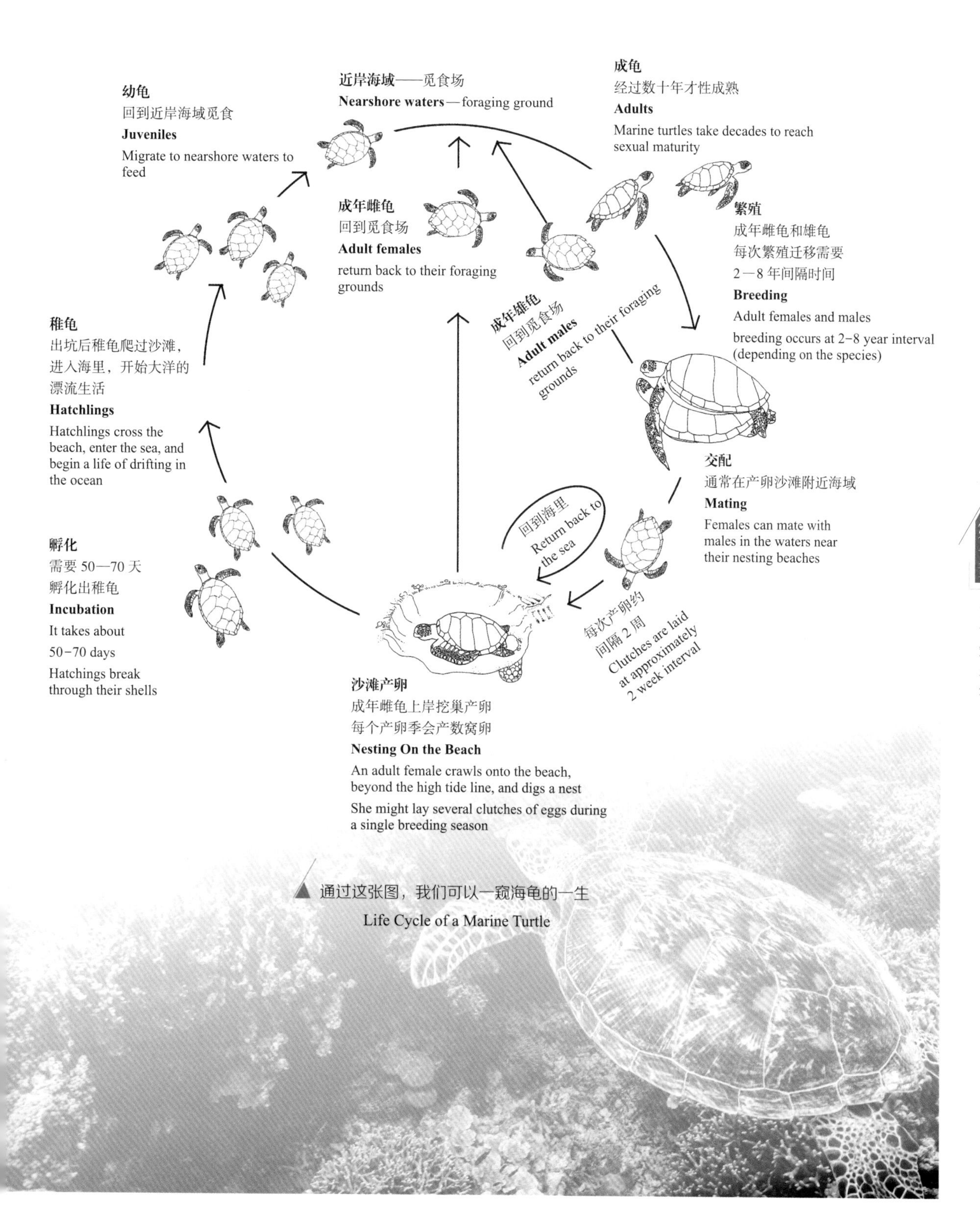

近岸海域——觅食场
Nearshore waters — foraging ground
幼龟
回到近岸海域觅食
Juveniles
Migrate to nearshore waters to feed
成龟
经过数十年才性成熟
Adults
Marine turtles take decades to reach sexual maturity
成年雌龟
回到觅食场
Adult females
return back to their foraging grounds
繁殖
成年雌龟和雄龟
每次繁殖迁移需要
2—8 年间隔时间
Breeding
Adult females and males
breeding occurs at 2–8 year interval (depending on the species)
成年雄龟
回到觅食场
Adult males
return back to their foraging grounds
稚龟
出坑后稚龟爬过沙滩，进入海里，开始大洋的漂流生活
Hatchlings
Hatchlings cross the beach, enter the sea, and begin a life of drifting in the ocean
交配
通常在产卵沙滩附近海域
Mating
Females can mate with males in the waters near their nesting beaches
回到海里
Return back to the sea
孵化
需要 50—70 天
孵化出稚龟
Incubation
It takes about
50–70 days
Hatchings break through their shells
每次产卵约
间隔 2 周
Clutches are laid at approximately 2 week interval
沙滩产卵
成年雌龟上岸挖巢产卵
每个产卵季会产数窝卵
Nesting On the Beach
An adult female crawls onto the beach, beyond the high tide line, and digs a nest
She might lay several clutches of eggs during a single breeding season

▲ 通过这张图，我们可以一窥海龟的一生

Life Cycle of a Marine Turtle

海龟生活在海里，但仍然需要浮出水面呼吸换气，获取氧气。不过，毫无疑问的是，它们是憋气的高手。海龟潜入水底的时间经常长达 1 小时。睡觉时，甚至可以几个小时不浮出水面呼吸。

海龟用肺呼吸，但其胸部并不能活动，采用的是一种吞吐气式的呼吸方式。待在水下时，它们则依靠泄殖腔壁丰富的毛细血管来吸收水中的氧气。

Marine turtles have lungs and therefore they have to emerge periodically in order to breathe. But, they can dive and stay underwater for up to an hour. However, they can stay underwater for several hours when they sleep.

Marine turtles breathe in air through external nares and pass to their lungs. They swallow air without thorax's movement. Perhaps more surprisingly, they can also obtain oxygen from the water via the cloaca (in the anus). It's full of blood vessels and by passing water over the cloaca, oxygen can be diffused into blood vessels.

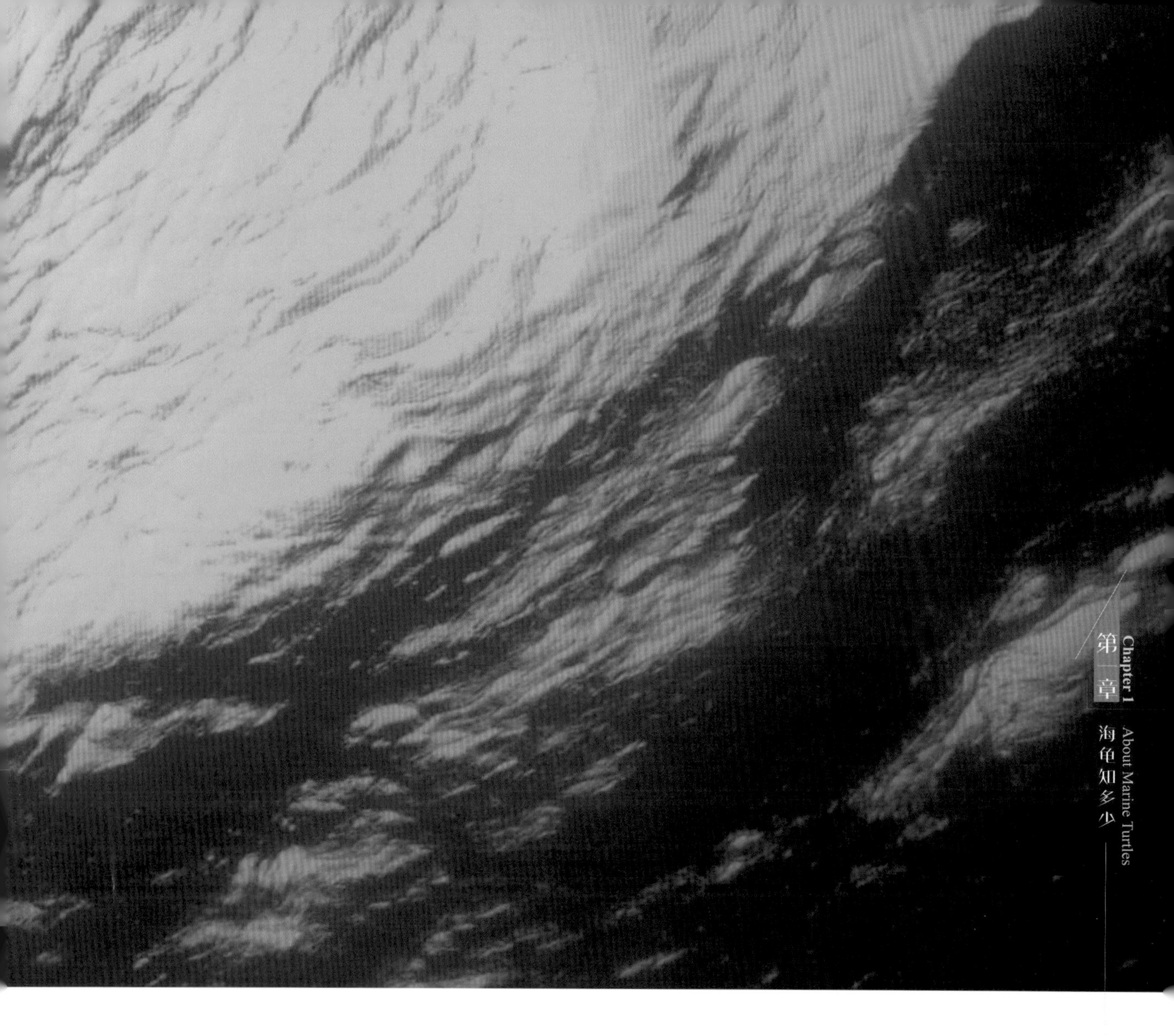

第一章 Chapter 1

海龟知多少 About Marine Turtles

一只绿海龟正在浮出水面换气
One green turtle comes up to surface to breath

© 杨位迪 Weidi Yang

海龟拥有大大的眼睛，它们的背甲和腹甲如装甲车一般坚固，保护它们免受天敌的攻击。但这样的身体结构也有劣势，海龟无法像一般龟类那样将头、颈和四肢缩回龟甲中。

幼龟，甚至非繁殖季节的成龟，都很难通过外形来辨别其性别。产卵季的雌龟尾巴粗短，而雄龟的尾巴较细长。

Marine turtles have big eyes, hard dorsal and abdominal shells and cannot retract their head, neck and limbs into the shell like some other turtles.

It is not easy to tell the gender of juvenile turtles or even adults outside the breeding season. During the breeding season, mature females can be distinguished by their short, thick tails, while mature males have long, thin tails.

海龟冷知识
Marine turtle trivia
海龟会翻身吗?
Can marine turtles get back on their feet when they are flipped over?
想知道答案吗，请在书中寻找
Want to know the answer? Please find in the book

不同种类的海龟有不同的食物偏好。比如，绿海龟的幼年时期主要为肉食性，成年后却转以海草和大型藻类为食；玳瑁喜食隐藏于珊瑚中的海绵；棱皮龟则偏爱水母。此外，它们还会吃海胆、贝类、甲壳类和鱼类。海龟没有牙齿，它们靠喙捕食。

Different species of marine turtles have different food preferences. Green turtles are mostly carnivorous when they are young but as they become adults, seaweed and macroalgae become their primary food source. Hawksbill turtles prefer sponges and leatherback turtles like jellyfish. In addition, marine turtles also feed on sea urchins, shellfish, crustaceans and fish. They do not have teeth, but rather they feed on prey using their beak.

海龟会流泪吗？

Can marine turtles cry?

想知道答案吗，请在书中寻找

Want to know the answer? Please find in the book

© 王思宇 Siyu Wang

第一章 海龟知多少

Chapter 1 About Marine Turtles

© Ace Wu

全世界共有 2 科 6 属 7 种海龟，均属于爬行纲的龟鳖目。其中，海龟科的 5 属 6 种是绿海龟、玳瑁、太平洋丽龟、红海龟、平背龟和肯普氏丽龟；棱皮龟科的 1 属 1 种是棱皮龟。在中国海域，生活着 5 种海龟，分别是绿海龟、玳瑁、太平洋丽龟、红海龟和棱皮龟，以绿海龟和玳瑁最为常见。

There are seven species of marine turtles in the world (split into 2 families and 6 genera). They are the green, hawksbill, olive ridley, loggerhead, flatback, Kemp's ridley and leatherback. Except for the Kemp's ridley and flatback, the five other species are found in Chinese waters with the green turtles and hawksbill turtles being the most common two species.

海龟放屁吗?

Can marine turtles fart?

想知道答案吗，请在书中寻找
Want to know the answer? Please find in the book

这 7 种海龟，我们该如何区分呢？

鉴定海龟种类主要看龟背的颜色、背甲上盾片的数目与排列方式，以及头部前额鳞的数目、排列方式等。下面我们就用以上方法来依次认识一下全球的 7 种海龟。

How can we tell different marine turtle species apart?

Marine turtles are identified mainly based on (1) the carapace color, (2) the number and arrangement of the vertebral and lateral scutes and (3) the number and arrangement of the prefrontal scales. Now let's identify the 7 species of marine turtles in the world using the aforementioned characteristics.

龟背颜色
Carapace color

· 成年太平洋丽龟的背甲是橄榄绿色

The carapace color of adult olive ridleys is olive green

椎盾和肋盾
Vertebral and lateral scutes

· 红海龟有 5 块椎盾、5 对肋盾

Loggerhead turtles have a carapace with 5 vertebral scutes and 5 pairs of lateral scutes

前额鳞数
Prefrontal scales

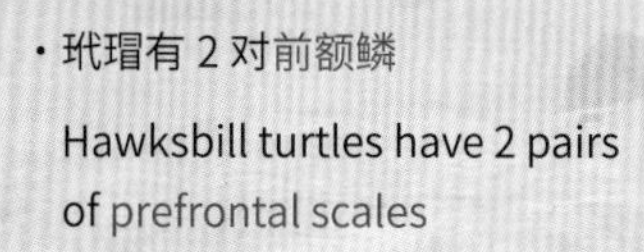

· 玳瑁有 2 对前额鳞

Hawksbill turtles have 2 pairs of prefrontal scales

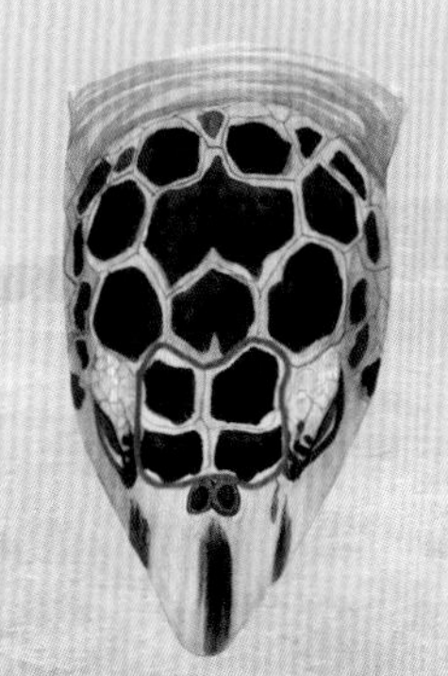

绿海龟 (*Chelonia mydas*)

绿海龟的头部有 1 对前额鳞，背甲上有 5 块椎盾、4 对肋盾，肋盾不与颈盾相连。绿海龟的稚龟背甲呈黑色，幼龟和成龟背甲呈棕色。

成年绿海龟以近岸浅水海域的海草和大型海藻为食，是海龟中唯一的素食者，因为皮肤下有一层黄绿色的脂肪而得名。虽然成年绿海龟主要吃素，但它们并非是“终身”的素食主义者，幼龟阶段主要以海洋无脊椎动物和鱼类为食，随着年龄的增长才逐渐偏向吃素。

绿海龟的产卵场分布于亚洲、大洋洲和美洲温暖海域的沙滩。我国的广东省惠东，香港特别行政区的南丫岛，台湾省的澎湖列岛、小琉球和兰屿，海南省的东沙群岛、西沙群岛、南沙群岛，都有绿海龟的产卵场。

Green turtles (*Chelonia mydas*)

Green turtles have one pair of prefrontal scales, 5 vertebral scutes and 4 pairs of lateral scutes with the first pair of lateral scutes not touching the nuchal scute (found directly behind the neck). The color of the carapace is black in hatchlings and brown in juveniles and adults.

Adult green turtles feed on seagrass near the shore and sometimes seaweed, making them the only vegetarians among the 7 marine turtle species. They are so named because their bodies have a yellow-green layer of fat under their skin, a result of their herbivorous diet. Although adult green turtles are primarily herbivores, juveniles feed on marine invertebrates and fish, and shift to plants as they grow older.

Green turtle nesting grounds are distributed widely on beaches in the warm waters of Asia, Oceania and the Americas. In China, the nesting grounds of green turtles are found in Huidong of Guangdong province, Lamma Island of Hong Kong, the Penghu Islands, Xiaoliuqiu and Lanyu around Taiwan province, and Dongsha Island, Xisha Islands (Paracel Islands) and Taiping Island in the South China Sea.

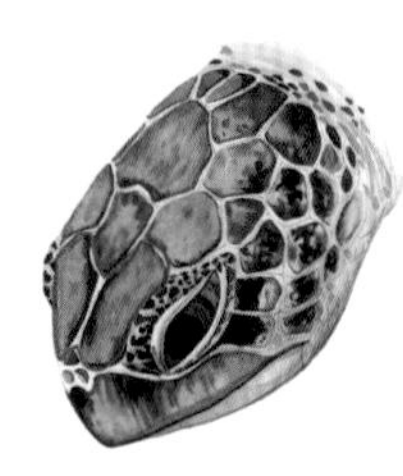

・前额鳞只有 1 对

Only 1 pair of prefrontal scales

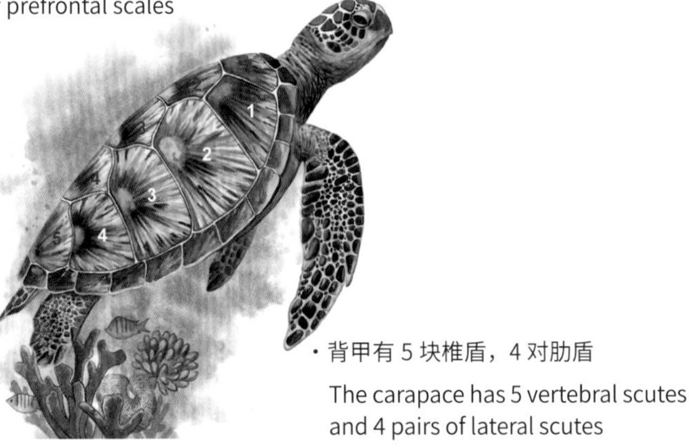

・背甲有 5 块椎盾，4 对肋盾

The carapace has 5 vertebral scutes and 4 pairs of lateral scutes

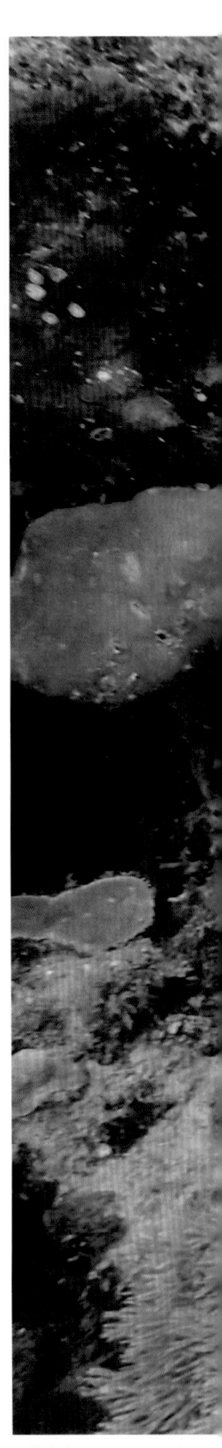

© 盖广生 Guangsheng Gai

一只绿海龟正在岩礁上寻找食物
A green turtle searching for food on the reef

▲ 一只正在悠闲“散步”的玳瑁
A hawksbill turtle going for a leisurely swim

玳瑁 (*Eretmochelys imbricata*)

玳瑁的头部有 2 对前额鳞，背甲上有 5 块椎盾、4 对肋盾，肋盾不与颈盾相连。背甲的盾片看上去就像屋顶上的瓦片叠加排列一样。成年玳瑁背甲上有漂亮的花纹，呈黄、黑和棕褐色的放射状图案。这样美丽的背甲被称为“海金”，给玳瑁带来美誉的同时也带来了灾难。人类觊觎这些美丽的图案，捕杀了无以计数的玳瑁。玳瑁跟绿海龟长得很相似，两者最大的区别在于：玳瑁的头部细长，有着尖锐的喙状嘴，而且上喙比下喙长，看上去像鹰嘴。玳瑁是凶猛的食肉动物，它们捕食鱼、虾、蟹、水母，不过，它们最喜欢的是珊瑚礁上的海绵。

玳瑁的产卵场分布广泛，在亚洲、大洋洲的澳大利亚和美国夏威夷的沙滩上均有玳瑁产卵的记录。我国是否有玳瑁的产卵场有待进一步证实。

Hawksbill turtles (*Eretmochelys imbricata*)

Hawksbill turtles have two pairs of prefrontal scales, 5 vertebral scutes and 4 pairs of lateral scutes with the first pair of lateral scutes not touching the nuchal scute. Their carapaces look like roof tiles with adults having scutes that are yellow, tan and brown, radiating to form a beautiful irregular pattern. It is this feature that makes the hawksbills beloved and also puts them in danger. Countless hawksbills have been hunted because humans covet their beautiful carapaces, utilizing them to make "tortoise shell" trinkets and jewelry found in markets around the world. Hawksbills have a long slender head and a sharp beak with a longer upper jaw, reminiscent of a hawk's mouth. They are aggressive carnivores, feeding on a variety of fish, shrimp, crabs and jellyfish; however, their favorite food are the sponges that grow on tropical coral reefs.

The nesting grounds of hawksbill turtles are widely distributed around the world, on beaches in the warm waters of Asia, Australia in Oceania and Hawaii in the United States. The nesting grounds of hawksbills in China merit further confirmation.

· 前额鳞有 2 对

2 pairs of prefrontal scales

· 背部有 5 块椎盾，4 块肋盾

The carapace has 5 vertebral scutes and 4 pairs of lateral scutes

太平洋丽龟 (*Lepidochelys olivacea*)

太平洋丽龟头部有 2 对前额鳞。背甲上有 6 块或以上椎盾、6—9 对肋盾，肋盾与颈盾相连。稚龟和幼龟背甲颜色为灰色，成年后才变为墨绿色。它们的背甲是心形的。太平洋丽龟食性很杂，包括鱼类、水母、甲壳类和软体动物。

太平洋丽龟的产卵场广泛分布在印度洋和大西洋，它们会在同一时间同一地点同时上岸产卵，群体数量甚至可达上万只，场面非常壮观。这道美丽的风景有个专门的名称，叫作“集体产卵”。太平洋丽龟的两个集体产卵场分别位于哥斯达黎加以及印度。我国没有太平洋丽龟的产卵场。

▲ 一只完成产卵的太平洋丽龟正在返回大海
An olive ridley sea turtle returning to the sea

Olive ridley turtles (*Lepidochelys olivacea*)

Olive ridley turtles have two pairs of prefrontal scales, at least 6 vertebral scutes and 6-9 pairs of lateral scutes with the first pair of lateral scutes touching the nuchal scute. The carapace is grey in hatchlings and juveniles and turns dark green and becomes heart-shaped after maturing. Olive ridley turtles have a varied diet, including fish, jellyfish, crustaceans and mollusks.

The nesting grounds of olive ridley turtles are widely distributed in the Indian and Atlantic Oceans. Hundreds to thousands of olive ridley turtles come ashore at the same time and in the same site to lay their eggs, creating a spectacular sight in nature. This beautiful sight has a special name: the "arribada". Two well-known arribada nesting grounds are found in Costa Rica and India. No olive ridley nesting grounds are found in China.

海龟放屁吗?
Can marine turtles fart?

可以。像人类一样，肠道消化食物时产生的气体需要通过肛门排出。

Yes, they can. Just like in humans, gases can be generated during food digestion in the gut of marine turtles and can be released through the anus.

© JHVEPhoto

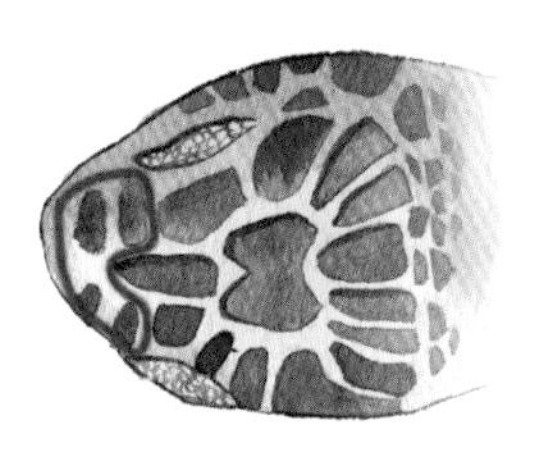

・前额鳞有 2 对

2 pairs of prefrontal scales

・背甲有至少 6 块椎盾、6－9 对肋盾

The carapace has at least 6 vertebral scutes and 6–9 pairs of lateral scutes

▲ 这只红海龟的背上藤壶和绿藻有点多，它会不会正在去“洗澡”的路上呢？
This loggerhead turtle has a lot of barnacles and green algae on its back. Is it on its way to a "cleaning station"?

© Naturepics_Li

红海龟 (*Caretta caretta*)

红海龟也叫“蠵(xī)龟”。红海龟的头部的前额鳞为2对，并且在2对前额鳞之间还有1块小鳞片。背甲上有5块椎盾、5对肋盾，肋盾与颈盾相连。红海龟可以很容易从它大而厚的头部和粗短的颈部加以辨识。成年红海龟体色为红褐色。红海龟通常捕食底栖无脊椎动物（如海绵）、头足类（如章鱼、乌贼）和鱼类。

红海龟的产卵场分布在印度洋、太平洋和大西洋。在太平洋的产卵场主要位于日本和澳大利亚。我国是否有红海龟的产卵场有待进一步证实。

Loggerhead turtles (*Caretta caretta*) (known as the “red sea turtle” in Chinese)

Loggerhead turtles have two pairs of prefrontal scales with another small scale in the middle, five vertebral scutes and five pairs of lateral scutes with the first pair of lateral scutes touching the nuchal scute. Loggerheads have a large, thick head and short, thick neck. Adult loggerheads have a reddish-brown body color. Loggerhead turtles usually feed on sponges, cephalopods and fish.

The nesting grounds of loggerhead turtles are distributed in the Indian, Pacific, and Atlantic Oceans; in the Pacific Ocean, the nesting grounds are mainly found in Japan and Australia. The nesting grounds of loggerhead turtles in China merit further confirmation.

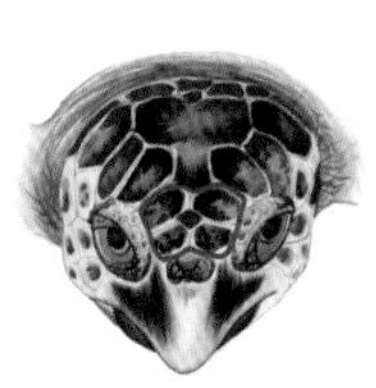

· 前额鳞有2对，且2对中间有1块小鳞片

5 (2 pairs and a small one in the middle) prefrontal scales

· 背甲有5块椎盾，5对肋盾

The carapace has 5 vertebral scutes and 5 pairs of lateral scutes

上岸产卵的棱皮龟
A leatherback turtle comes to shore to lay eggs

© Rawlinson_Photography

棱皮龟 (*Dermochelys coriacea*)

棱皮龟是唯一一种背部覆盖着厚厚的皮层而非盾片的海龟。棱皮龟的背部有 7 条突起的纵脊，头部无鳞片。成年棱皮龟龟皮为深灰色或黑色，并带有许多白色斑点。棱皮龟是地球上现存最大的海龟，身长可达 2.5 米，也是游泳能力最强的海龟。它们最喜欢的食物是水母，因此获得了一个别称，叫“吃水母的潜艇”。

棱皮龟的产卵场分布在太平洋、大西洋以及印度洋。我国没有棱皮龟的产卵场，在我国海域出现的棱皮龟主要来自东南亚的产卵群体。

Leatherback turtles (*Dermochelys coriacea*)

Leatherback turtles are the only one out of the 7 marine turtle species to have a layer of thick, rubbery skin instead of a bony carapace. They have 7 ridges on their backs and no prefrontal scales. The body is dark grey or black with pale spots. Leatherbacks are the largest of the marine turtle species (up to 2.5 meters in body length) and the best swimmers. Their favorite food is jellyfish, so they have a nickname “the submarine eating jellyfish” in Chinese. The nesting grounds of leatherback turtles are distributed in the Pacific, Atlantic and Indian Oceans. No nesting grounds of leatherback turtles are found in China; the leatherbacks occurring in Chinese waters are mainly from Southeast Asian nesting populations.

· 头部无鳞片
背甲无盾片，有 7 条纵脊

No scales on the head
No bony carapace, with 7 long ridges

© SheraleeS

平背龟 (*Natator depressus*)

平背龟的背甲比其他海龟都要平坦，故而得此名。平背龟仅生活在澳大利亚北部、印度尼西亚和巴布亚新几内亚附近的浅水海湾，尤其喜欢河口和珊瑚礁区。平背龟以水母、蛤类、虾类为食，也吃海草等。虽然成年平背龟的个头跟其他海龟比起来不算大，但其稚龟的体型却是所有海龟稚龟中最大的。

平背龟的产卵场仅在澳大利亚北部。

© 北京市企业家环保基金会（SEE Foundation）

Flatback turtles (*Natator depressus*)

Flatback turtles are so named because their carapaces are flatter than the other marine turtle species. They are only found in the shallow bays around Indonesia, Australia and Papua New Guinea, particularly in estuaries and coral reefs. Flatback turtles feed on jellyfish, clams, shrimp and seaweed. Although adult flatbacks are not very large among marine turtles, the hatchlings are the largest of all marine turtle hatchlings.

The nesting grounds for flatback turtles can only be found in northern Australia.

肯普氏丽龟 (*Lepidochelys kempii*)

肯普氏丽龟是世界上最小的海龟，也是世界上最稀少的海龟。它们喜欢水母、螃蟹、蛤类等食物。肯普氏丽龟仅分布在墨西哥湾和美国西海岸。

肯普氏丽龟的产卵场主要分布在墨西哥湾。受产卵场退化、海洋污染和人类捕杀等威胁影响，肯普氏丽龟种群数量锐减，几乎到了灭绝的边缘。近年来，尽管墨西哥政府建立了严格的保护体系，国际组织也给予大力支持，肯普氏丽龟的种群数量仍未有恢复的迹象。

Kemp's ridley turtles (*Lepidochelys kempii*)

Kemp's ridley turtles are the smallest and rarest marine turtle in the world. They feed on jellyfish, crabs and clams. Kemp's ridley turtles are only distributed in the Gulf of Mexico and along the Atlantic coast of the United States.

The nesting grounds of Kemp's ridley turtles are found only along the Gulf of Mexico. Due to illegal egg harvesting, habitat deterioration, pollution, natural catastrophes and fishery bycatch, the population size of the Kemp's ridley declined dramatically. Although the Mexican government has instituted strict laws to protect their nesting grounds in recent years with support from international organizations, there is still no significant recovery of Kemp's ridley populations.

© Evgenia76

第二章
Chapter 2

海龟的轮回
Marine Turtles' Reincarnation: from Generation to Generation

我们的世界是“永恒轮回”的。我们生命的每 一秒都有无数次的重复。
——米兰·昆德拉《不能承受的生命之轻》

“If every second of our lives recurs an infinite number of times, we are nailed to eternity as Jesus Christ was nailed to the cross.”
(*The Unbearable Lightness of Being* by Milan Kundera)

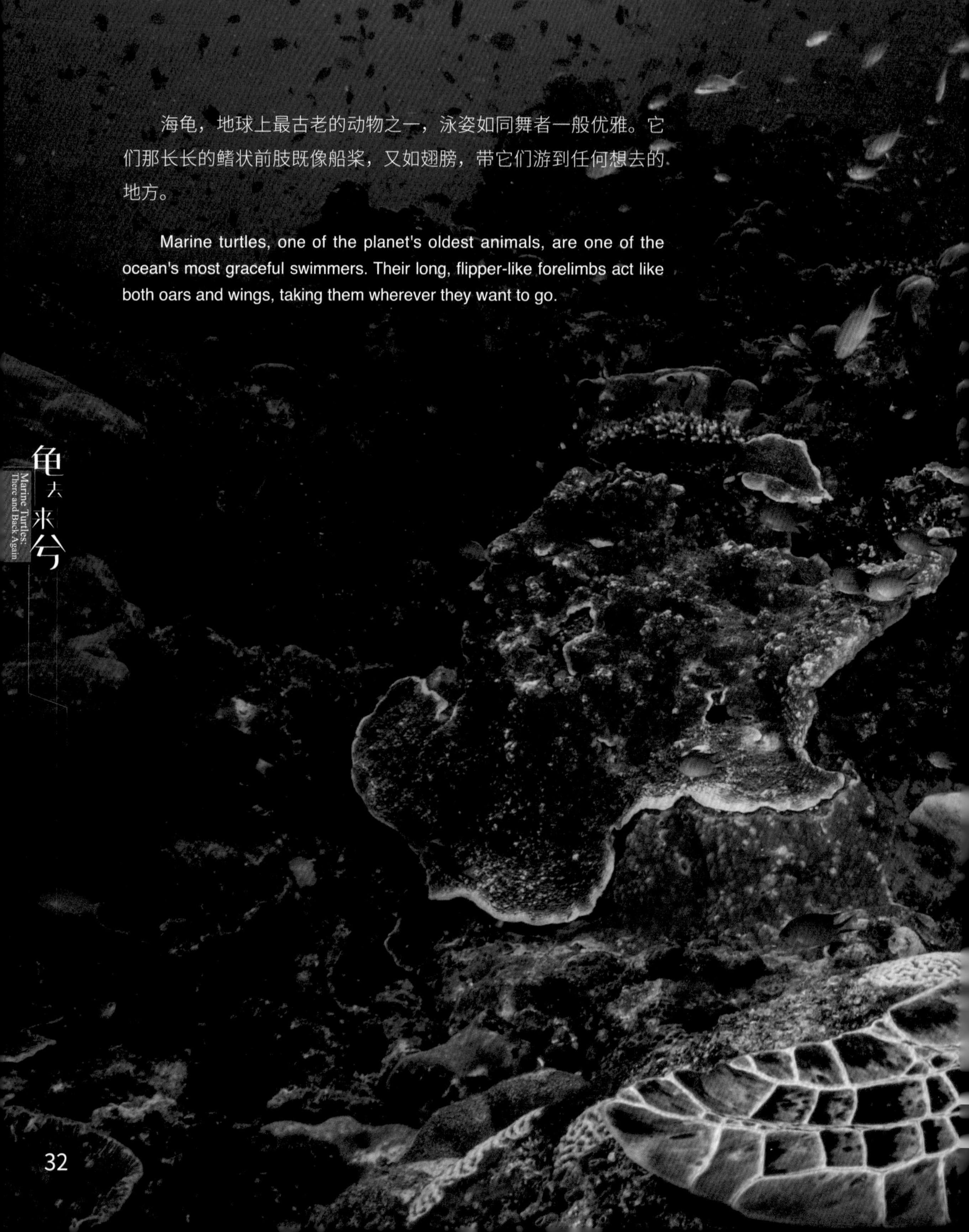

海龟，地球上最古老的动物之一，泳姿如同舞者一般优雅。它们那长长的鳍状前肢既像船桨，又如翅膀，带它们游到任何想去的地方。

Marine turtles, one of the planet's oldest animals, are one of the ocean's most graceful swimmers. Their long, flipper-like forelimbs act like both oars and wings, taking them wherever they want to go.

第二章 海龟的轮回

Chapter 2 Marine Turtles' Reincarnation: from Generation to Generation

如此自由自在，那海龟又会游去哪儿呢？ 它们会游去食物丰盛的地方。在海水温度下降后，海龟会迁移到水温较高的水域来抵御寒冷，它们甚至还会进行海底“冬眠”，让自己的新陈代谢降低。

Where do marine turtles go? Marine turtles migrate to areas with plentiful food. When the temperature drops too low, they migrate to warmer waters and may even undergo a period of brumation to slow their metabolisms.

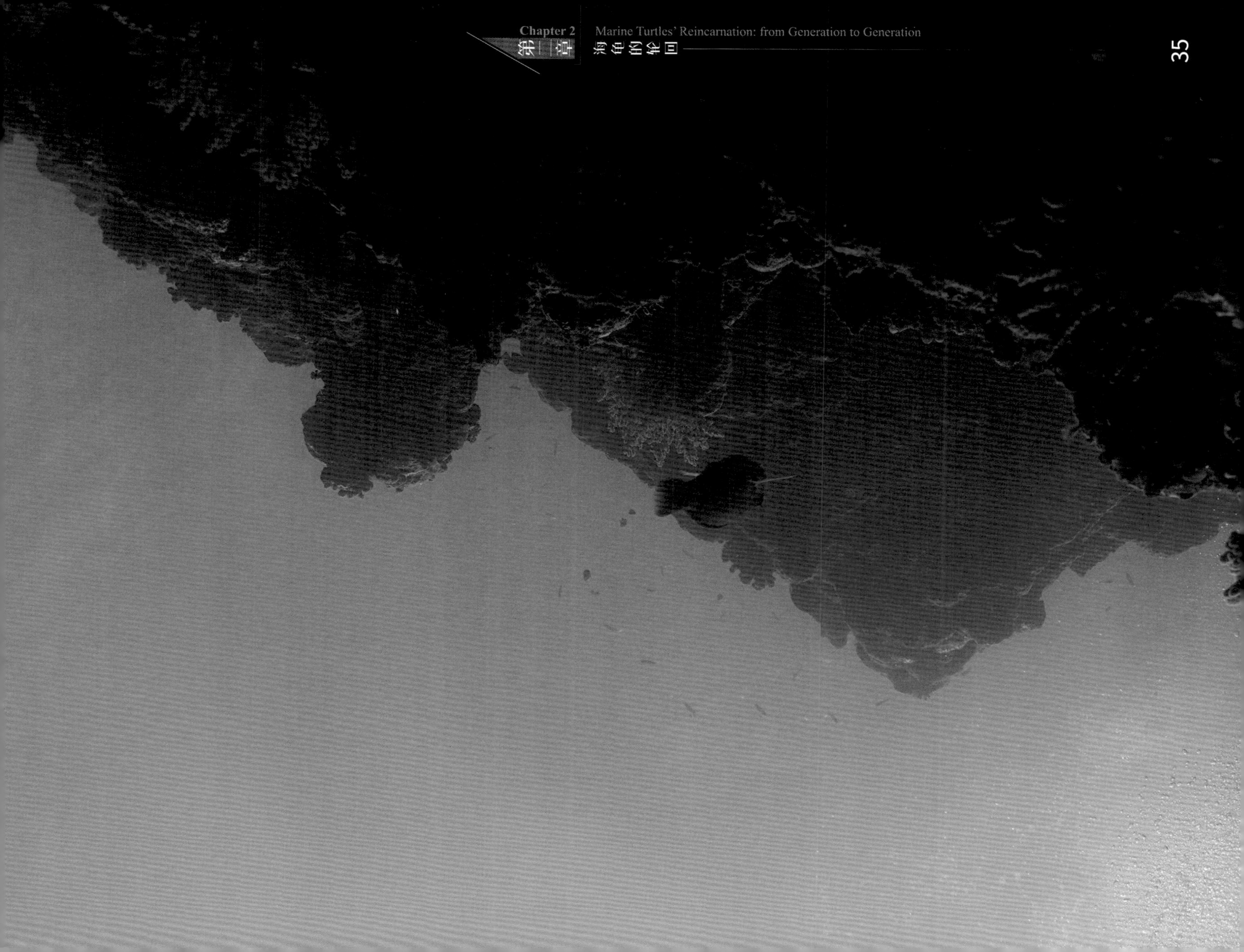

第二章 Chapter 2

海龟的轮回

Marine Turtles' Reincarnation: from Generation to Generation

海龟一生中最重要的游泳旅程，就是繁殖季进行的长距离迁徙。性成熟的成年雄性和雌性海龟会从四面八方的觅食场回到自己的出生地，繁衍下一代。产卵场和觅食场之间的距离可能是上百或上千千米。

The most significant migration for marine turtles occurs during the breeding season, when mature females and males may travel hundreds or thousands of kilometers from different foraging grounds back to their birth places in order to mate and lay their eggs.

海龟会流泪吗?
Can marine turtles cry?

会。但是海龟流泪不是因为伤心，是用来分泌盐，来维持体内水分和盐的平衡。

Yes, they cry. However, it is not because they are sad. Marine turtles secrete tears to excrete extra salt out of the body to maintain their body's homeostasis.

© 王洁 Jie Wang

▲好舒服的海龟“巴士”
Such a comfortable marine turtle “bus”

▲这张图有一个关于海龟的趣事（想知道答案吗？请在本书中寻找）
An interesting story about the picture (Want to know the answer? Please find in the book)

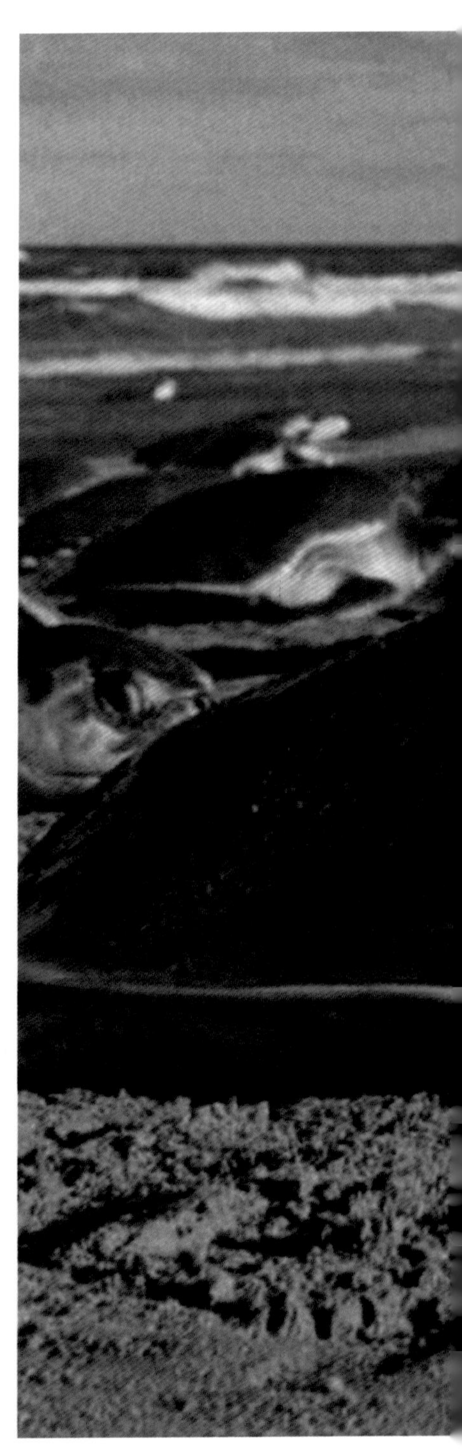

海龟产卵高峰多在夏季。在此期间，成年雄龟和雌龟都会返回自己的出生地，在靠近产卵场的海域交配 。雄龟通常会比雌龟更早到达出生地附近海域，因为海龟的交配原则是先到先得。雌龟可以与数头雄龟交配，并将各只雄龟的精子储存在泄殖腔内，等到卵子分批成熟后进行受精。也正是因为如此，雌龟在一个产卵季节可多次上岸产卵。而一次产卵季节过后，间隔 2－8 年，雌龟才会再次产卵，成年雄龟可每年交配。另外，成年雄龟和雌龟也有可能在进行繁殖迁移的途中就交配了。

The peak nesting seasons of marine turtles usually occur during summer and females go ashore to lay their eggs when the water is warm and calm. Mature males usually arrive at the waters close to their birth beaches earlier than mature females because the mating principle is first come, first served. A female can mate with more than one male and store the sperm in their cloaca in order to fertilize different clutches of eggs. A female can lay several clutches of eggs in one nesting season. While adult males can mate again the following year, females have a 2-8 year interval before mating again. In addition, the mature females and males can also mate while migrating to their birth places.

▲蔚为壮观的太平洋丽龟集群产卵
An olive ridley "arribada" – a spectacular sight

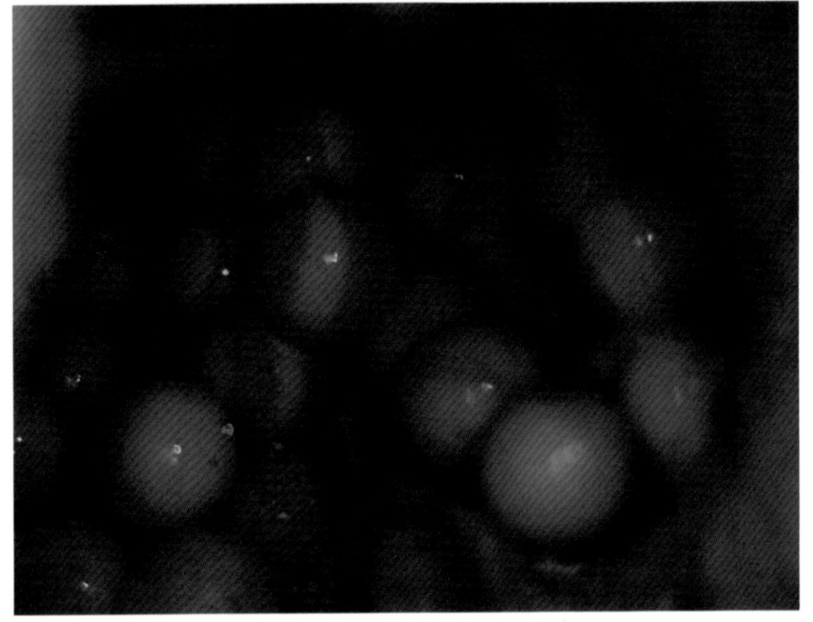

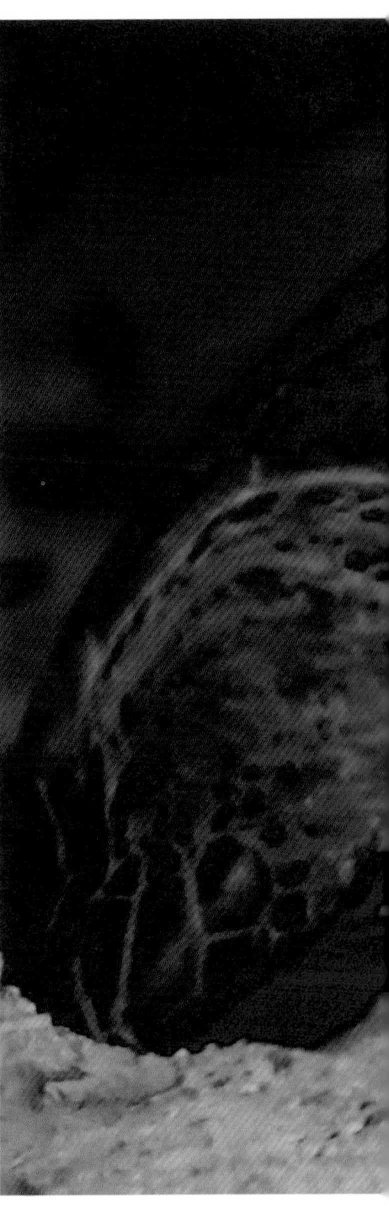

交配后的雌龟通常会在太阳落山后前往人烟稀少和有遮蔽物、处于高潮线位置以上的沙滩挖掘巢穴，产卵完毕后再将巢穴用沙盖好才离开。

Female marine turtles usually nest after sunset, searching for sheltered and undisturbed beaches and spots above the high tide line to lay their eggs. They will cover the nests with sand before returning to the sea.

© 三沙市海洋保护区管理局 Sansa Marine Nature Reserve Management Bureau

一个产卵季，不同的龟产卵的数量会不同。有些雌龟甚至可以产多达 8 窝卵。每窝 80—170 枚，每窝间隔时间约为 2 周。通常，一枚龟卵的大小跟一个乒乓球差不多。

In one nesting season, the number of clutches that a female can lay varies by marine turtle species. Some females can lay up to 8 clutches of eggs. A clutch contains 80–170 eggs. The interval between each clutch is approximately 2 weeks. The eggs look like ping pong balls.

02

到沙滩了，雌龟开始用前肢大力地把沙子朝身后推出去，挖出一个跟自己体型差不多大小的坑穴。不过，这通常并不是它们最终产卵的蛋坑。它们会不停地尝试新的位置，挖啊挖，直到满意为止。

Once she has chosen her spot, the female will use her front flippers to dig a hole (body pit) about the same size as her own body. The first hole may not be the final one, and the female will keep moving on the beach, digging holes until she finds the right spot.

03

挖好满意的坑穴后，就要在其中进一
出蛋坑了。挖蛋坑时，雌龟用的是后肢，
沙子是朝前方扬出去的。雌龟的身体后部
断地抬起放下且左右摇摆，还不时停下来休

She will then further excavate an
chamber using her rear flippers. Her bod
move up and down, swaying from left to
sometimes stopping as she pauses to res

01

雌龟缓慢地浮出海面，爬一会儿，停一会儿，为的是不断观察周边是否存在潜在的威胁。此时，任何轻微的声音或者亮光，以及它们感觉到的任何可能的危险，都会让雌龟立刻返回大海，放弃这次产卵。

A female turtle will come out of the water slowly, constantly pausing and scanning for any sign of danger as she crawls up the beach. They will give up nesting and return to sea if disturbed by any noise or artificial light.

06

最后，该返回海里去了……

Finally, it's time for her to return to ocean……

05

产卵结束了。雌龟先用后肢小心地刨沙埋住蛋坑，然后再用前肢大力扬沙掩饰蛋坑四周的环境。

After she is done laying her eggs, she will carefully cover them with sand using her rear flippers. She will then use her front flippers to refill the body pit and then fling sand all around in order to disguise the nest.

04

终于，产卵正式开始啦。雌龟的产卵过程非常安静， 偶尔传出有节奏的喘息声和轻摆后肢的声音。如果说产卵之前任何风吹草动都会惊吓到雌龟的话，那么产卵开始之后，则无论外界多大的动静它们都不会分神。即使这样，人们最好不要打搅它们。

Finally, she begins to lay her eggs and once she starts, she will continue until all her eggs are laid. Although it is unlikely that she will stop laying once she has started, if disturbed, it is possible and therefore they should never be disturbed while they are nesting.

夏天艳阳照耀着沙滩，温暖着蛋坑里的海龟卵。大约 2 个月，稚龟们将来到这个对它们而言全新的世界。破壳时，稚龟会用鼻前一个小而尖的点啄破蛋壳，这个特殊的部位在稚龟脱壳之后就会消失。

和大多数动物不同，决定稚龟性别的，不是染色体，而是孵化时蛋坑里的温度，较高的温度会孵化出更多的雌龟。同一窝的稚龟会在几天内陆续孵化，先出壳的稚龟会在蛋坑内等待并呼唤未出壳的弟弟妹妹，直到达到一定数量，再相互合作着离开蛋坑。在爬出蛋坑的过程中，顶上的沙会落入下方空的蛋壳中，形成稚龟可以往上爬的阶梯。不出几天，稚龟就可以全部爬出来了。

Hatchlings will emerge in about 2 months (the incubation time can vary with temperature). Hatchlings have a small, sharp beak in the front of the nostrils which is used to break the egg shell, and then disappears after hatching.

The incubation temperature in the nest determines the hatchlings' genders, with higher temperatures resulting in more females. Hatchlings that hatched first will call and wait for their siblings to break their egg shells. When the time is right, they will burst out of the sand en masse. Hatchlings from the same clutch will usually take a few days to come out of the egg chamber completely.

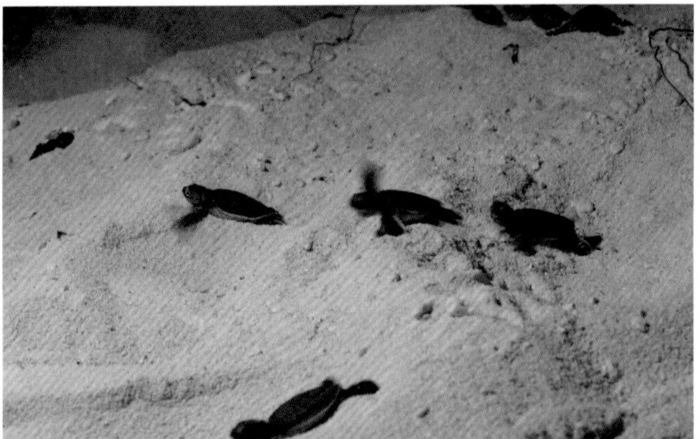

在朦胧的夜色或晨曦中，数以百计的稚龟不计困难、不畏危险，摇摆着小小的四肢，坚定不移地离开坑穴，爬过沙滩，直奔大海。幸存下来的稚龟开始了幼龟的生活，随海流从近岸迁移到大洋，躲避在海藻团中以浮游动物为食。我们对这个时期的海龟知之甚少，称为海龟“迷失的岁月”。

Hatchlings will crawl toward the ocean before sunrise and after sunset when the beaches are not too hot. The hatchlings swing their front and rear flippers as they make their mad dash toward the water. Those that make it from the nest to the water will then swim with all their might as they are still not safe; the nearshore waters are teeming with predators. Those lucky enough to survive will get picked up by currents and taken farther out to the open ocean, where they will hide in clumps of floating macroalgae and feed on zooplankton and start growing into juveniles. From this point on, not much is known about the lives of these animals. This is when the “lost years” begins.

◀ 稚龟们会在朦胧夜色或者晨曦中奔向大海。这两个时段的光照不强烈，稚龟们不会被晒伤

Hatchlings crawl toward the ocean usually before sunrise and after sunset when the beaches are not too hot and they won’t get burnt

经过几年的大洋生活，幼龟有了更强的自主捕食和躲避天敌的能力。它们漂洋过海，迁移数千千米，向近岸海域靠近。它们一边捕食，一边继续成长。十几年后，幼龟终于脱去稚气，迈入成年期。这也意味着，在下一个产卵季到来时，它们将像它们的父母那样，重回自己的故土，“结婚生子”。

Juvenile turtles spend years in the open ocean, honing their feeding skills and becoming more adept at avoiding predators. Juvenile turtles will migrate hundreds or thousands of kilometers across the ocean, moving closer to nearshore waters to feed. After a decade or more, they are finally adults. This means that when the next breeding season arrives, they will return to the beaches of their births, to mate, nest and continue the legacy.

很多人都好奇：海龟是怎么做到如此准确地洄游到自己多年以前出生的那片沙滩的呢？据科学家们推测，海龟能够通过地球磁场和太阳及其他星体的位置来辨别方向并识别出生地的地形，最终到达某个特定的目的地。

或许，这正是神奇的生命轮回。

How do marine turtles manage to migrate so accurately to the beaches of their births? According to research, marine turtles are able to orient using the Earth's magnetic field and the position of the sun and stars to find their way home.

Perhaps this is the miracle created by the miraculous cycle of life.

从太空俯视地球，会发现地球大部分区域被海水覆盖，因此，我们的星球也被称为“蓝色星球”。海洋是地球上最大的生态系统，所有的海洋生物构成了一张巨大的食物网和生态网，它们相互依赖，又相互制约，达到生态系统的平衡。

Our planet is covered largely by the ocean, and so it’s also known as the “blue planet”. The ocean is the largest habitat on Earth and all marine organisms form a vast marine food web and ecological network. They depend on and restrict each other to achieve a balanced ecosystem.

虽然人们把地球上的海洋划分为几个大洋和一些附属海，但事实上，它们之间并没有相互隔离，而是形成一个连续的整体。 海洋生物也是如此，它们同样是一个整体。在海洋食物网中，任何生物的存在，都有其“平衡生态”的重要作用。海龟也不例外，它们对海洋物质循环和能量流动而言十分重要，在维持珊瑚礁、海草床等关键生态系统的健康稳定方面发挥着不可替代的作用。

Although we divide the ocean into several large oceans and some subsidiary seas, in fact, they are not isolated from each other and are a continuous whole. In the vast marine food web, every organism plays an important role to maintain ecosystem balance. Marine turtles are important in maintaining the cycle of materials and energy flux, and the health and stability of key ecosystems such as coral reefs and seagrass meadows.

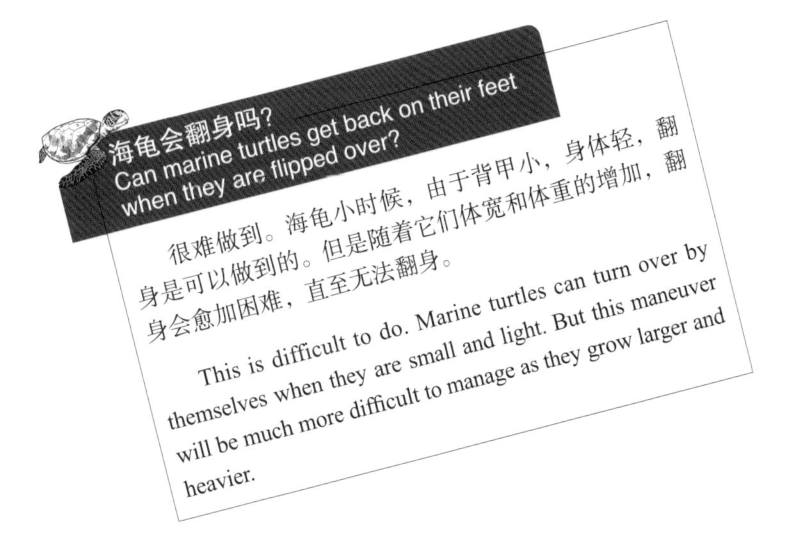

海龟会翻身吗?
Can marine turtles get back on their feet when they are flipped over?
很难做到。海龟小时候，由于背甲小，身体轻，翻身是可以做到的。但是随着它们体宽和体重的增加，翻身会愈加困难，直至无法翻身。
This is difficult to do. Marine turtles can turn over by themselves when they are small and light. But this maneuver will be much more difficult to manage as they grow larger and heavier.

绿海龟啃食海草，可以帮助修剪海草，海草在被啃食后可以生长的更好。生长旺盛的海草可以稳固海底的沙砾，为其他海洋生物提供优良的栖息地、避难所、育幼区和食物来源。

Green turtles feed on seagrass, thus maintaining the health and growth of seagrass meadows. Healthy seagrass creates a stable sea floor and acts as a habitat, refuge, nursery and food source for other marine organisms.

一只绿海龟正在津津有味地吃着海草
A green turtle eating seagrass

© 问海自然保护中心 Hi Ocean Nature Conservation Center

珊瑚礁形成速度缓慢，珊瑚礁上的海绵繁殖过快会抑制珊瑚的生长。海龟对海绵的捕食，不仅有效地控制了海绵的数量，也维持了珊瑚的健康状况，依赖珊瑚礁生活的其他海洋生物也将得到更好的生存和发展。

Coral reefs grow slowly and the outbreak of sponges will inhibit the growth of corals. Those marine turtles that feed on sponges can effectively control their numbers, thus maintaining the health of coral reefs, so that other marine organisms dependent on coral reefs can have a better chance for survival.

海龟冷知识
Marine turtle trivia
海龟有指头吗?
Do marine turtles have fingers?
想知道答案吗，请在书中寻找
Want to know the answer? Please find in the book

© 吴江 Jiang WU

海龟喜食水母，可谓是水母的天敌。维护海龟种群数量的稳定，在一定程度上抑制了水母的数量骤增，防止了水母的大量爆发。虽然有些水母有一定的食用价值，但是营养价值都不高。更糟糕的是，水母多以小鱼小虾为食，因此水母的爆发，对水生生物资源的影响是不容忽视的。我们必须承认，海龟的存在造福了海洋生态系统。

Jellyfish are marine turtles' favorite food. Maintaining a healthy and sustainable population of marine turtles can control jellyfish numbers and prevent jellyfish blooms. Although some jellyfish species can be consumed by humans, they have low nutritional value. Even worse, jellyfish feed on juvenile fish and small shrimps; therefore, a jellyfish bloom will impact fishery production significantly. We have to admit that the existence of marine turtles is a benefit to the entire marine ecosystem.

海龟冷知识
Marine turtle trivia
海龟会生病吗?
Can marine turtles get sick?
想知道答案吗，请在书中寻找
Want to know the answer? Please find in the book

第二章 Chapter 2

海龟的轮回

Marine Turtles' Reincarnation: from Generation to Generation

▲拼命爬向大海的稚龟们，它们会成功吗？
Hatchlings struggling towards the ocean. Will they succeed?

© 张婷 Ting Zhang

海龟对于海洋来说如此不可或缺，但它们一生命运坎坷，充满了危险。刚出生的稚龟背甲是软的，加上个头比较小，所以在奔向大海的沙滩上，很容易被海鸟、螃蟹和其他动物（比如蜥蜴）吃掉。即使是成功投入大海怀抱的稚龟，以及幼龟和成龟也面临许多敌人，如鲨鱼、海狼鱼甚至鲸类都可能吃掉它们。

Marine turtles are essential for a healthy ocean. However, once they come into the world, their lives are fraught with danger. Hatchlings are easy prey for birds, crabs and other animals e.g. (lizards) while they make their way to the water. Those hatchlings, who manage to make it to the sea and grow will still face many predators, including large sharks, baracuda and whales during their grow out.

38 页图片答案 Answer of Page 38 picture

当雌龟在海中遇到“胡搅蛮缠”的雄龟，甩都甩不掉的时候，雌龟会选择背着雄龟一起上岸，如你在图中所见。离水上岸后，雄龟很快会在重力的作用下滑下来，这样雌龟就有机会甩掉这只雄龟。

When a female turtle encounters a male one, which she doesn't like and can't get rid of it in the sea, sometimes she will crawl up the beach with the male on her back, as you see in the picture. On the beach, the male will soon slide down because of gravity so that the female has a chance to get rid of the male.

© Jolo Diaz Pexels

未成功孵化的海龟卵会成为沙滩植被的养分来源，植被如果生长茂盛，又可以更好地固定沙砾、保护沙滩。部分龟卵和沙滩上刚孵化出来的稚龟、大洋中的幼龟，也是一些海鸟、螃蟹和鱼类的食物来源，成年海龟也会遭到一些大型捕食者，如鲨鱼和虎鲸的捕猎——这些都利于维持海洋食物网的稳定。

Unhatched eggs provide a source of nutrition for beach vegetation while the eggs, hatchlings and juvenile turtles are also food sources for seabirds, crabs and fish. Even adult marine turtles can be a food source for large predators, such as sharks and killer whales. These maintain the stability of the marine food web.

你知道吗？
Did you know?
海龟产下的卵中，大约有 20% 不会孵化成功。
Approximately 20% of marine turtle eggs will not hatch successfully.

海龟冷知识
Marine turtle trivia
海龟会打架吗？
Do marine turtles fight with each other?
想知道答案吗，请在书中寻找
Want to know the answer? Please find in the book

第三章
Chapter 3

海龟的危机
The Crises of Marine Turtles

当人类欢呼对自然的胜利之时，也就是自然对人类惩罚的开始。——恩格斯

"Let us not, however, flatter ourselves overmuch on account of our human victories over nature. For each such victory nature takes its revenge on us." by Friedrich Engels

虽然海龟会受到凶猛的鲨鱼、海狼鱼和鲸类的攻击，但它们往往可以凭借坚硬的龟甲和高超的游泳技术躲开大部分的捕猎者。成年海龟需要面对的最大的天敌，是——人类！人类认为海龟的卵、肉、背甲具有药用价值、食用价值和文化价值，所以海龟成了人类疯狂捕杀的对象。捕杀者会滥挖龟卵，还会用海龟的背甲做成装饰品售卖，这些行为都使得海龟的数量急剧下降。受高额利润的驱使，海龟非法捕捞和非法贸易在全球屡禁不止。

Even though marine turtles are attacked by aggressive sharks, barracuda and killer whales, they can run away from most predators because of their hard carapace and swimming skill. The biggest enemy for adult marine turtles are human beings. People hunt them for their eggs, meat and shells for food, medicine, to make trinkets and for various other cultural reasons both globally and regionally. Such activities have resulted in the dramatic decline of marine turtle populations. Driven by high profit, illegal hunting and trade happens around the world.

海洋生态系统是地球生态系统的重要一环。海洋生物的大量消亡，会为整个地球的生态系统增添一分脆弱。那么，同样生活在地球上的人类，真的可以独善其身吗?

The marine ecosystem is critically important to the Earth ecosystem. The extinction of too many marine lives will increase the vulnerability of the entire Earth ecosystem. We all live on the same Earth and cannot survive alone.

在我国海域，海龟的生存面临着同样的威胁，包括误捕、非法贸易、栖息地丧失、海洋污染等。近海渔业捕捞渔具多种多样，包括拖网、定置网、流刺网等，都可能误捕、误伤海龟。被误捕的海龟中，以绿海龟最为常见。这些海龟因不小心钻进正在进行捕捞作业的渔网中，而被渔民捞上船。

2019 年的一项研究数据显示：33 只通过定位追踪方式研究（放归）的海龟中，其中 5 只来自罚没，4 只在上岸产卵后安装定位追踪放归，2 只来自在海上漂浮被发现救护，3 只来自人工繁育，19 只来自渔民误捕（拖网和定置网）救护，可见误捕的威胁之大。

Marine turtles in China face the same threats as they do around the world, such as illegal egg harvest, bycatch, illegal trade, habitat loss and pollution. Fishing gear in nearshore waters are diverse, including trawl nets, set nets and gill nets. All these can accidentally trap marine turtles. Among marine turtle bycatch, green turtles are the most common species seen.

According to one review paper in 2019, of 33 released marine turtles in a satellite tracking study in Chinese waters, 5 were confiscated from illegal trade, 4 were tagged while nesting on the beach, 2 were rescued at sea, 3 were artificially bred and 19 were found as bycatch in fishing nets (trawl and set nets), thus indicating fishery bycatch as their biggest threat to survival.

海龟会打架吗？
Do marine turtles fight with each other?

几乎不会。海龟非常绅士，它们之间几乎不会有身体上的冲突。但是，如果我们将海龟饲养在很小的水体里，海龟会在高度紧张压力下相互伤害。

Rarely. Marine turtles are gentle creatures who rarely have physical conflict with each other. However, when they are kept in small pools or tanks, they become highly stressed and fights can occur.

© 胡全君 Quanjun Hu

▲ 浙江舟山误捕的一只红海龟，渔民在检查无恙后放归

A loggerhead turtle accidentally captured by a fishing vessel in Zhoushan, Zhejiang Province, China. It was released after checking for injuries

© 北京市水生野生动植物救护中心 Beijing Aquatic Wildlife Rescue Center

有些海龟是被误捕，有些海龟则是被专门捕获用来进行非法贸易的，比如被捕获来作为“镇店之宝”或龟甲被加工成工艺品出售。

Some marine turtles are accidentally caught as bycatch and some are captured for the illegal trade.

玳瑁因为有漂亮的背甲花纹，是非法捕捞和非法贸易中的头号目标。人类残忍地把玳瑁的背甲割下来，做成各种玳瑁制品，尤其是工艺品。尽管中国法律和《濒危野生动植物物种国际贸易公约》都明令禁止海龟及其制品买卖，但由于海龟制品在市场上很受欢迎，许多不法分子选择铤而走险，为牟利而滥捕海龟。

Hawksbill turtles are the most targeted species because of their beautiful carapace. People make various products out of the shell. Although the trade of marine turtles and their products are prohibited by law in China and under CITES Appendix I, hunting still occurs, because demand and illegal trade still exist.

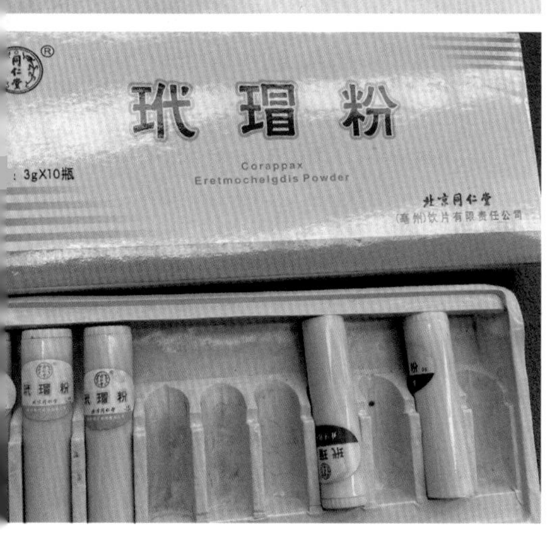

栖息地丧失严重威胁着我国海龟的生存。随着海岸带的开发，历史上一些天然的海龟产卵场，比如广东和海南沿岸的许多沙滩，被人类活动占据，现在已经基本上看不到绿海龟在这些区域上岸产卵了。同时，科学家证实沿岸灯光污染对海龟上岸产卵和稚龟入海都有负面影响。产卵的海龟对于灯光极其敏感，只要有明显的灯光，就不会上岸产卵。刚孵化的稚龟，本能会朝着光亮的地方爬去（即海平面上的月亮），由于沿岸灯光干扰，导致稚龟向相反的方向爬，远离大海。

Habitat loss is a major threat to the survival of marine turtles in China. With coastal development, some natural marine turtle nesting grounds have been overrun by human activities such as seen along the many beaches in Guangdong and Hainan Provinces. Coastal light pollution also has a negative impact on marine turtles' nesting and hatchlings making it to the sea. Marine turtles are extremely sensitive to light and will not come ashore to lay eggs if there is visible light. Hatchlings will head toward the brightest point on the horizon and this should be the moon above the sea, however, light pollution causes them to go the wrong direction, i.e. away from the sea.

科研工作者在研究海龟产卵时使用特殊的红光，尽量避免强光对于产卵海龟的影响

Researchers use special red lights when studying nesting marine turtles so as to avoid affecting them as they lay eggs

第三章 Chapter 3

海龟的危机 The Crises of Marine Turtles

▲海龟吞食塑料袋
A marine turtle ingesting a plastic bag

▲在海底发现被废弃渔网缠绕而死去的海龟碎片
Fragments of dead marine turtles in discarded fishing nets

海洋垃圾，尤其是塑料垃圾正在威胁海龟的生存。由于塑料袋的外形很像水母，海龟会把其误认为是自己的食物。海龟吞食塑料后堵塞消化道造成消化不良，严重的会致死。海洋中布设的定置张网、作业的拖网以及其他废弃渔具也成了海龟的陷阱。被渔网缠住的海龟，会因不能及时浮出水面呼吸而窒息死亡。

Marine debris, especially plastic, also threatens the survival of marine turtles as it is possible they confuse plastic bags for food such as jellyfish. The plastics cannot be digested, which may result in death. Set nets, trawl nets and discarded fishing nets can unintentionally trap marine turtles, resulting in suffocation and death if they are not rescued in time.

▲工作人员在水下打捞废弃渔具
Salvaging discarded fishing gear underwater

海洋垃圾的泛滥和被污染的水环境严重影响海龟的健康。曾有研究发现，比起其他海龟，绿海龟更易在吞食海草时误食塑料，特别是形状狭长、绿色及黑色、更像海草的塑料。英国埃克塞特大学和塞浦路斯海龟保护协会的科学家们检查了塞浦路斯海滩上的海龟尸体，发现所有海龟的胃肠道里都有塑料存在，最多的一只胃肠道里有183 片塑料。产卵场的海滩垃圾不仅会干扰雌海龟的筑巢活动和巢址选择，导致筑巢失败或改变巢穴分布格局，也会阻碍刚孵化的稚龟向大海爬行，增加其被捕食的风险。

Marine debris and pollution are seriously affecting the health of marine turtles. It has been found that green turtles are more likely to eat plastic while feeding on seagrass, especially thin, green and black pieces of plastic which look like seagrass. In all of the dead marine turtles found on Cypriot beaches, plastic was found in the stomachs and intestines with 183 pieces of plastic found in one turtle. Debris on nesting grounds not only interferes with female nesting activities and site selection, leading to nesting failure, but also prevents hatchlings from crawling to the sea and increases the risk of predation on the beach.

认识海洋垃圾

About marine debris

联合国环境规划署将海洋垃圾定义为：被丢弃、弃置或遗弃在海洋以及沿海环境里的任何持久性、人造的或经加工的固体物质。海洋垃圾包括人为制造或使用并被故意弃置于海洋、河流或海滩，或由河流、污水、暴风雨、大风直接携带入海的固体废物，以及恶劣天气条件下意外遗失的渔具、货物等。常见的海洋垃圾材质有塑料、合成橡胶、合成纤维、玻璃、金属等。一部分海洋垃圾停留在海滩上，称为“海滩垃圾”；一部分海洋垃圾则漂浮在海面或沉入海底，称为“海面漂浮垃圾”和“海底垃圾”。仅是太平洋上的海洋垃圾就达 300 多万平方千米，比一个印度还要大！

Marine debris is defined by the United Nations Environment Programme (UNEP) as “any persistent, manufactured, processed, or solid material discarded, disposed of, or abandoned in the marine and coastal environment”. Common marine debris includes plastics, synthetic rubber, synthetic fibers, glass and metals. Some debris stays on the beach, while others will float in the water and some others will sink to the sea floor.

微塑料的定义

尺寸小于 5 毫米的塑料纤维、颗粒或者薄膜被学界定义为微塑料，并认为对海洋和水生生物具有危害。

Definition of microplastics

Microplastics are small plastic pieces (including fibers, particles and membranes) less than 5 mm in size, which can be harmful to our ocean and aquatic organisms.

▲ 海滩垃圾
Beach trash

◀ 海面漂浮垃圾
Floating trash

© 杨子弈 Ziyi Yang

▲ 海底垃圾
Seafloor trash

© Lucas Meneses

海滩垃圾主要是塑料袋、塑料瓶、烟头、聚苯乙烯塑料泡沫快餐盒、渔网和玻璃瓶等；海面漂浮垃圾主要是塑料袋、漂浮木块、浮标和塑料瓶等；海底垃圾主要是玻璃瓶、塑料袋、饮料罐和渔网等。

Beach debris consists mainly of plastic bags, plastic bottles, cigarette butts, polystyrene plastic foam, fishing nets and glass bottles. Floating debris consists mainly of plastic bags, buoys and plastic bottles. Debris found on the seafloor consists mainly of glass bottles, plastic bags, other drink cans and fishing nets.

海洋中最大的塑料垃圾是废弃的渔网，有的废弃渔网长达几千米，人们把它们叫作“鬼网”。在洋流的作用下，这些渔网绞在一起，就成了众多海洋动物的“死亡陷阱”。 每年都会有数千只海龟、海豹、海狮和海豚等被渔网缠住、淹死。联合国粮农组织的一项研究显示，全球海洋垃圾中 10%—30% 为废弃渔网。

Discarded fishing nets, sometimes miles long, are also known as 'ghost nets'. A study by the Food and Agriculture Organization of the United Nations (FAO) shows that 10%–30% of global marine debris consists of derelict and discarded fishing nets.

© 张晶 Jing Zhang

▲ 被渔网缠绕的海龟
A marine turtle trapped by fishing net

© Greenpeace / Marco Care

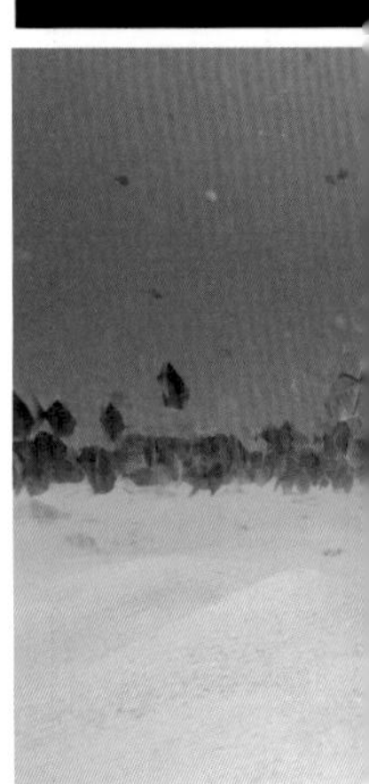

塑料制品在动物体内无法消化和分解，误食后会引起胃部不适，造成行动异常、繁殖能力下降，甚至导致死亡。不可小视的还有微塑料。游荡的微塑料很容易被浮游动物、滤食者（如贻贝）这样位于食物链底端的生物吃下，然后，通过食物链传递，使得微塑料在不同营养级的海洋生物体内传递、累积，导致海洋生物生病或死亡，甚至借由餐桌上的海产品进入人类的身体。据澳大利亚纽卡斯尔大学的一项研究显示，塑料污染已经侵入人类体内，全球人均每周摄入近 5 克的微塑料（约等于一张信用卡或银行卡所用的塑料）。海洋生物生病或死亡，会导致海洋生态系统的紊乱，最终影响整个地球的生态系统。届时，每一个人都将承担无法想象的后果……

Plastic products cannot be digested or broken down in an animal's digestive tract and can cause stomach upset, abnormal movement, reduced fertility and even death. Microplastics cannot be ignored either. Microplastics can be easily consumed by zooplankton and filter feeders such as mussels at the bottom of the food chain. These organisms transfer and accumulate the microplastics up through the food chain, possibly resulting in sickness or death to those organisms further up. They can even enter human bodies through seafood consumption. According to one study, each person consumes nearly 5 grams of microplastics (roughly the amount of plastic used in a credit or debit card) a week. Unhealthy marine organisms will lead to the disruption of the marine ecosystem, which will eventually influence the entire Earth ecosystem. At that point everyone will bear the unimaginable consequences……

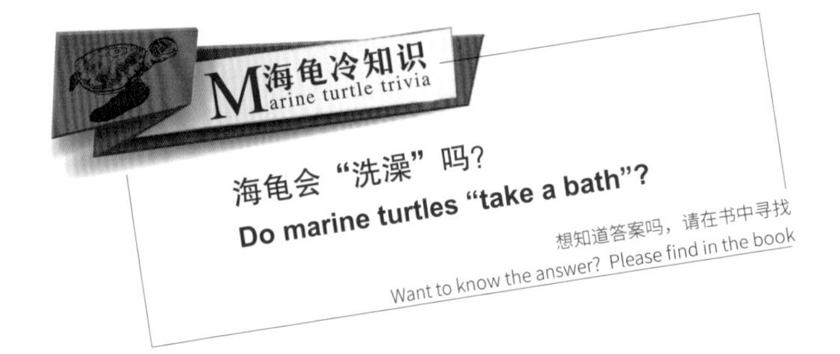

海龟冷知识
Marine turtle trivia
海龟会“洗澡”吗？
Do marine turtles “take a bath”?
想知道答案吗，请在书中寻找
Want to know the answer? Please find in the book

真实的故事 A true story

如前所述，我国西沙群岛七连屿北岛的西北部沙滩是绿海龟最重要的产卵场之一。海龟蛋坑附近的沙土中常常含有大量塑料瓶，刚出巢的幼龟也出现过被塑料渔网缠绕的现象。 这里的海滩塑料垃圾有来自邻近的越南、马来西亚、菲律宾、印度尼西亚等东南亚国家的，很可能是这些垃圾进入海洋后，在夏季西南季风和洋流的作用下被输送至此。

The northwestern beach on the North Island of Qilianyu cluster, Xisha Islands, is one of the most important nesting grounds for green turtles in China.Yet large amounts of plastic bottles can be found on the beach. The debris, which is likely transported by the southwest monsoon typhoons and ocean currents in summer, comes from Southeast Asian countries such as Vietnam, Malaysia, the Philippines and Indonesia.

海龟会“洗澡”吗？
Do marine turtles "take a bath"?

会。藤壶等固着生物会附着在海龟背上，并侵蚀它们的背甲。海龟会利用珊瑚石当“搓澡巾”，把身上的藤壶蹭掉。

Yes, they do. Barnacles can attach on the back of marine turtles, which will damage their carapaces. Marine turtles will rap against corals to get rid of barnacles.

▲ 沙滩上的塑料瓶
Plastic bottles on the beach

▲ 稚龟与渔网
Green turtle hatchlings and fishing nets

© 张婷 Ting Zhang

产卵季海龟在沙滩上留下的痕迹。但人类活动正在让它们不断失去可以产卵的沙滩

Marine turtles'trails in nesting season. But the turtles are losting their nesting ground because of human activities

海滩道路、酒店及娱乐设施照明

Roads, hotels and entertainment facilities light the beach

除此之外，海上工程建设以及海岸线开发也会伤害海龟。它们会干扰和阻拦海龟的繁殖迁徙或破坏其产卵场。如果不能回到出生地产卵，雌性海龟不得不将卵产到海里，这样的卵将不可能孵化。在种种因素的作用下，全球七种海龟均受到生存威胁。

Coastal development destroys nesting beaches, effectively putting a roadblock up in the reproductive cycle. Marine turtles show high fidelity to their natal beaches and we do not know if mature females will find new beaches to lay their eggs. Under the various threats aforementioned, all seven marine turtle species are threatened.

海龟有指头吗？ Do marine turtles have fingers?

有。很难相信海龟也是“五指动物”吧！海龟每肢有5根指（趾）头，只是被一整片皮肤覆盖住变成了“鳍状肢”。但是，海龟的指甲是可以被看见的，其数量还是区分海龟种类的重要依据。

Yes, they do. It is hard to believe that marine turtles also have five fingers on each limb, isn't it? Their 5 fingers are covered by a whole piece of skin which forms a flipper. However, we can see their claws (finger nails) on their flippers, and the number of claws is a feature for species identification.

此外还有——气候变化的影响：全球气候变暖和海平面上升也对海龟的生存有着重大影响。一方面海平面上升不断吞噬着一些岛礁和沙滩，导致位于高潮线以上的适合海龟产卵的沙滩不断缩小。另外一方面，孵化温度决定所孵化海龟的性别。研究表明，如果海龟孵化蛋坑的平均温度在 28.8°C左右，海龟的雌雄性别比例接近 1∶1；如果高于 30.5°C孵化出的海龟全部为雌性，而低于 28°C 孵化出的海龟全部为雄性。因此，日渐升高的全球温度正在打破海龟雌雄比例的平衡，对海龟种群结构的深远影响不可小觑。

Furthermore, global warming and sea level rise also impact the survival of marine turtles. On the one hand, sea level rise continues to swallow up islands and beaches, resulting in a shrinking of marine turtle nesting beaches. On the other, incubation temperature determines the sex of hatchlings. Studies have shown that when the average nest incubation temperature is around 28.8°C, the sex ratio of the hatchlings is close to 1:1. When the average temperature is above 30.5°C, hatchings are all females, while below 28 °C they all are males. Global warming therefore changes the 1:1 sex ratio which will impact the overall population structure.

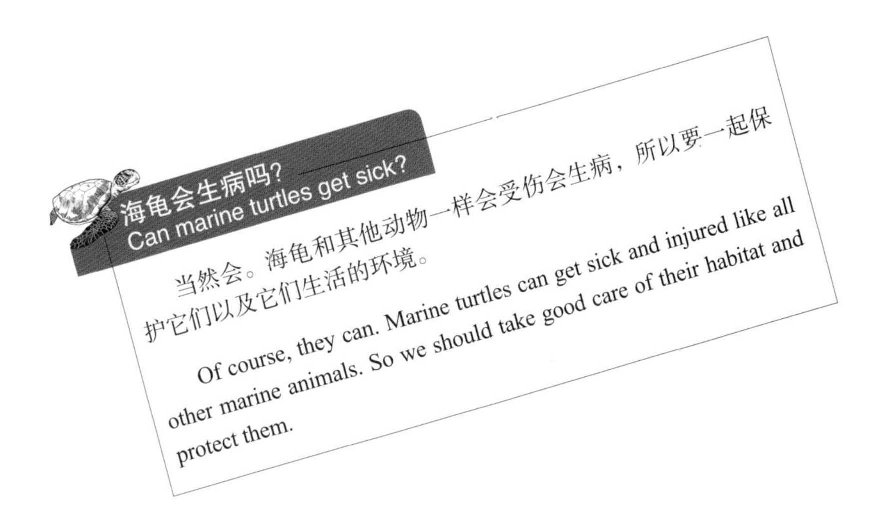

海龟会生病吗?
Can marine turtles get sick?
当然会。海龟和其他动物一样会受伤会生病，所以要一起保护它们以及它们生活的环境。
Of course, they can. Marine turtles can get sick and injured like all other marine animals. So we should take good care of their habitat and protect them.

© Quang Nguyen vinh

渔船、快艇、邮轮、货轮等船只在航行时，可能会由于速度过快碰撞到海龟而导致海龟受伤，旋转的螺旋桨也可能对海龟造成伤害。

Fishing vessels, speedboats, cruise ships and cargo vessels can cause injuries to marine turtles when accidental collisions occur, and spinning propellers can also cause damage to marine turtles.

第四章
Chapter 4

拯救海龟大作战
Let's Act to Save Marine Turtles

人的良心要求人去帮助一切能够帮助的有生之物，并免于去伤害任何生命；除非人能遵从这种良知，否则人就不能说是有道德的。——阿尔贝特·施韦泽《动物世界》

"The purpose of human life is to serve, and to show compassion and the will to help others. Only then have we ourselves become true human beings." (*The Animal World* by Albert Schweitzer)

在自然状态下，只有千分之一的海龟才能够从出生顺利长到性成熟。因此，每一只海龟都是生命的奇迹，都值得人类认真守护。

In nature, only one out of every thousand marine turtles survive to reach sexual maturity. So every marine turtle is a miracle of life and deserves to be carefully protected by us.

20 世纪 70 年代，在美国东南部海域，由于拖网捕虾等捕捞作业方式导致的海龟误捕已经成为海龟种群持续繁衍的重要威胁。为此，美国开始探索如何解决该问题，以使得被列入美国《濒危物种法案》的海龟得到有效保护。经过数十年的渔具改良实践和充分的评估后，最终研发出了海龟逃生装置（简称 TED）。研究表明，设计良好、安装得当、后期维护良好的 TED 可以使约 97% 误入拖网的海龟逃生。

In the 1970s, marine turtle bycatch by shrimp trawlers in the southeastern United States became a significant threat to the population. In response, the United States began to explore possible ways to address the problem so that the marine turtles, listed under the US *Endangered Species Act*, could be effectively protected. Through decades of fishing gear modification and evaluation, the Turtle Excluder Device (TED) was created. Studies have shown that a well-designed, properly installed and maintained TED can prevent almost all marine turtles (97%) that enter a trawl net from becoming trapped.

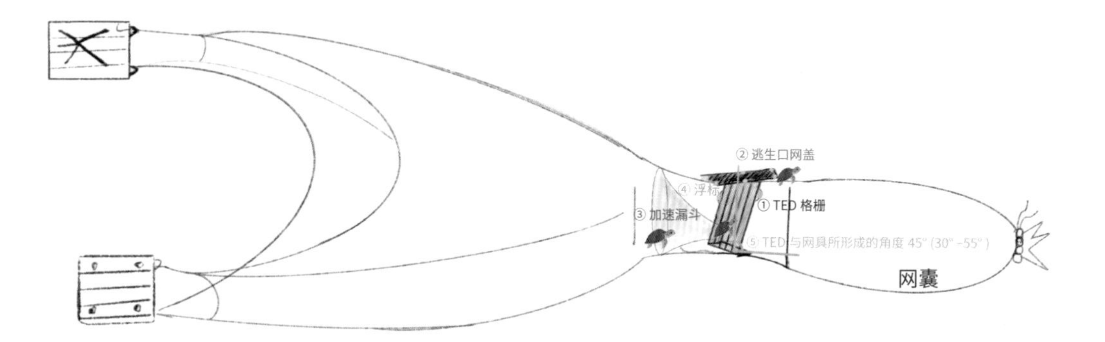

▲ 在双拖网作业中，在网具底部安装的海龟逃生装置（TED）示意图

The Turtle Excluder Device (TED) installed at the bottom of a trawl net

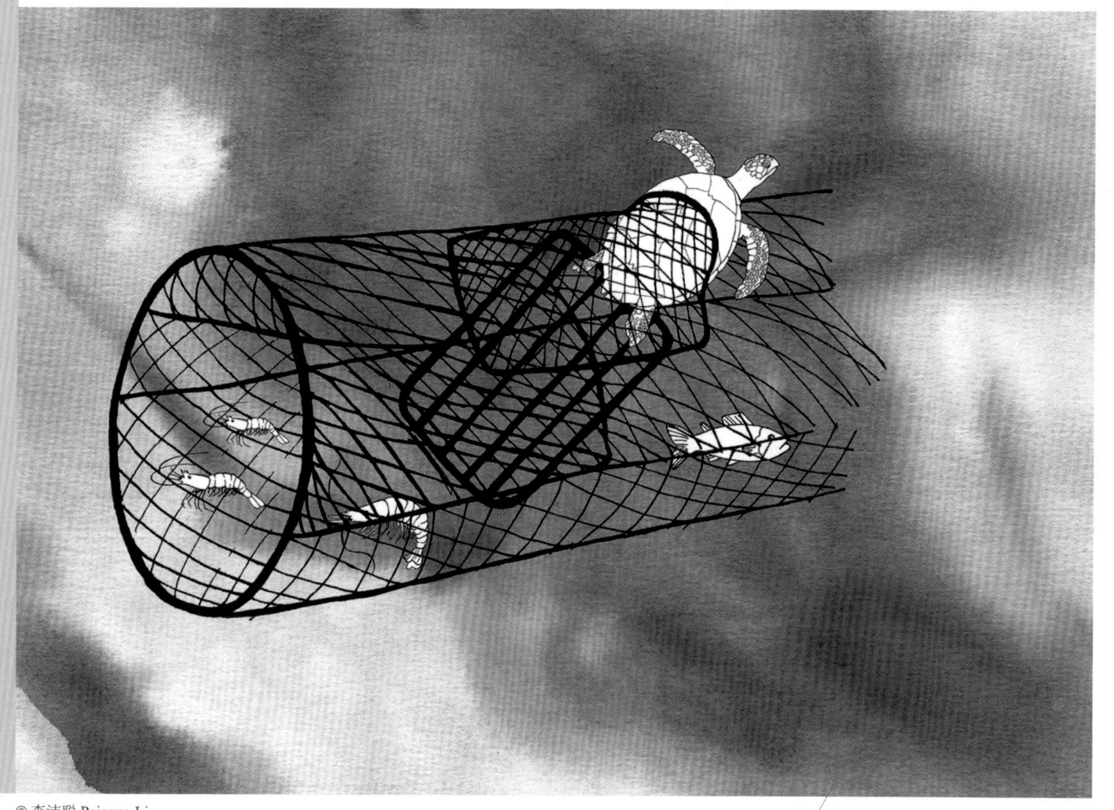

▲ 海龟成功从 TED 逃脱示意图

A turtle escapes through a TED

龟在中国人的心目中有着很特别的意义。不管是古代还是现代，龟都被视为解危消灾、避难降福的吉祥、长寿之物。海龟更是被视为神龟，甚至在一些寺庙中还会被供奉。

Turtles are significant creatures to Chinese people. They have been regarded as a symbol of auspiciousness and longevity in ancient and modern worlds, and are believed to drive away misfortune and bring blessings. Marine turtles are even regarded as divine and consecrated in some temples.

海龟不仅受到各种国际公约的监管，也受到中国国内法律的保护。《中华人民共和国野生动物保护法》规定，严格保护珍贵濒危野生动物，任何未经批准捕捉、人工繁育、出售购买利用海龟及其制品的行为，将被追究相应的法律责任。《中华人民共和国刑法》第 341 条规定，非法猎捕、杀害或收购、运输、出售国家重点保护的珍贵、濒危野生动物及其制品的，处五年以下有期徒刑或者拘役，并处罚金；情节严重的，处五年以上十年以下有期徒刑，并处罚金。

2021 年，新修订的《国家重点保护野生动物名录》将中国海域生活的五种海龟（红海龟、绿海龟、玳瑁、太平洋丽龟和棱皮龟）全部列为国家一级保护野生动物。

Marine turtles are protected not only under international conventions, but also by laws in China. Under the *Wildlife Protection Law of the People's Republic of China*, unauthorized capture, rearing, sale and purchase of marine turtles and their products will be prosecuted. According to Article 341 of the *Criminal Law of the People's Republic of China*, anyone who illegally hunts, kills, acquires, transports or sells precious or endangered wild animals and their products will be sentenced to up to ten years imprisonment and fined, depending on the circumstances.

In 2021, the five species of marine turtles (the green, the loggerhead, the hawksbill, the olive ridley and the leatherback) found in Chinese waters were officially upgraded to the category I protected animals at the national level.

第四章 Chapter 4

拯救海龟大作战

Let's Act to Save Marine Turtles

真实的故事 A true story

2017 年 4 月 19 日，广东省湛江市徐闻县渔政大队在海安港查获被大小搭配放在 5 个箱子里的 10 只海龟，之后将其运往徐闻珊瑚礁国家级自然保护区管理局救护池，并在救护观察一段时间之后，放归大海。

On 19th April, 2017, 10 marine turtles in five boxes were seized by Xuwen County Marine Police, Zhanjiang City, Guangdong Province. They were kept at the Xuwen Coral Reef National Nature Reserve and released after recovery.

你知道吗？
Did you know?

海龟放归也是需要条件的，最好选择有沙滩的海域放生，并且海水清澈。

When releasing marine turtles, a sandy beach and good quality water is preferred.

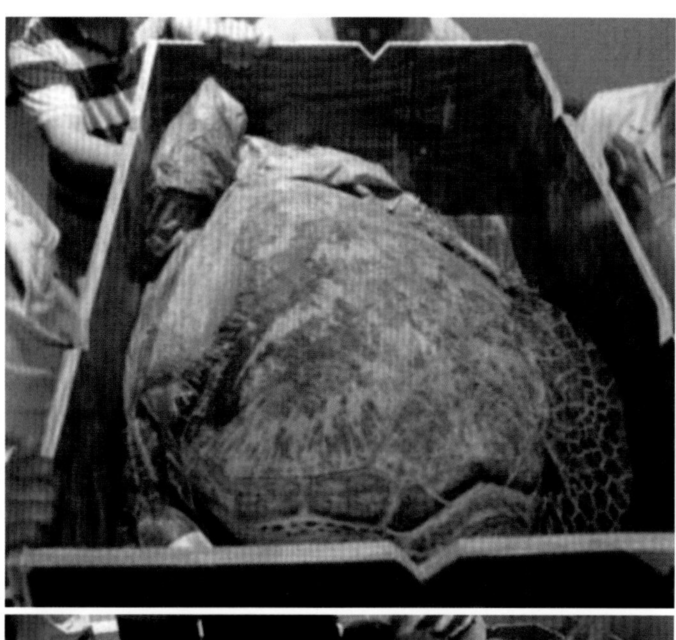

被救护的绿海龟

A rescued green turtle

根据卫星定位追踪发现——海龟在我国出现最频繁的近岸海域包括福建、广东、广西和海南等附近海域。台湾海峡、琼州海峡和吕宋海峡都是海龟的主要洄游廊道，海龟从这里游往日本、越南和菲律宾。

中国现有的海龟产卵场则主要集中在中国南海西沙群岛的一些岛屿沙滩，比如七连屿以及甘泉岛、晋卿岛等。广东惠东、香港南丫以及台湾万安也有我国现存的为数不多的近岸绿海龟产卵场。

According to satellite tracking data, marine turtles in Chinese waters are found mostly in the coastal waters of Fujian, Guangdong, Guangxi and Hainan Provinces. The Taiwan, Qiongzhou and Luzon Straits are the main migratory corridors through where marine turtles migrate to Japan, Vietnam and the Philippines.

Today, the main nesting grounds of green turtles in China are on the beaches of the Xisha Islands in the South China Sea, such as the Qilianyu cluster and Ganquan Island and Jinqing Island. Huidong of Guangdong Province, Lamma of Hong Kong and Wangan of Taiwan are the extant few nesting grounds.

你知道吗？

卫星定位追踪是研究海龟生活习性、洄游路线以及产卵场和觅食场之间洄游廊道的主要方法。科学家们通过一种特殊胶水将追踪器黏合到海龟的背甲上，在放归大海之前，设定好追踪器发出信号的时间等参数，并启动追踪器。除了记录海龟的位置外，追踪器也可以记录水深、水温等信息。当海龟浮出水面呼吸时，追踪器上发射的信号会被工作卫星接收并传输回地面接收站，再转送至处理中心。如此一来，海龟所在位置就会被记录下来。科学家们将一定时段内下载的地理位置信息绘制成海龟的洄游路径图，从而确定被追踪海龟的活动区域。

Did you know?

Satellite tracking is one of the main tools used to study marine turtle behaviors, migratory routes and the migratory corridors between nesting and feeding grounds. Researchers can attach a transmitter to the carapace which then transmits the locations of these marine turtles to the satellite and back to ground stations. Researchers then process location data to map the migratory routes of marine turtles.

真实的故事 A true story

这只幼龟是 2016 年在我国福建厦门海域放归的。但在不到一周的时间，人们就在我国台湾金门料罗湾海滩上发现了它的尸体。经过解剖，幼龟的体内有尼龙材质的渔网线和塑料软管等异物，但是直接的死因未知。

This juvenile green turtle was released in Xiamen City, Fujian Province in 2016. Within a week after its release, its body was found on the beach at Jinmen Island, Taiwan. While the cause of death remains unknown, nylon rope and plastic hoses discovered in its digestive tract upon necropsy may have been a contributing factor.

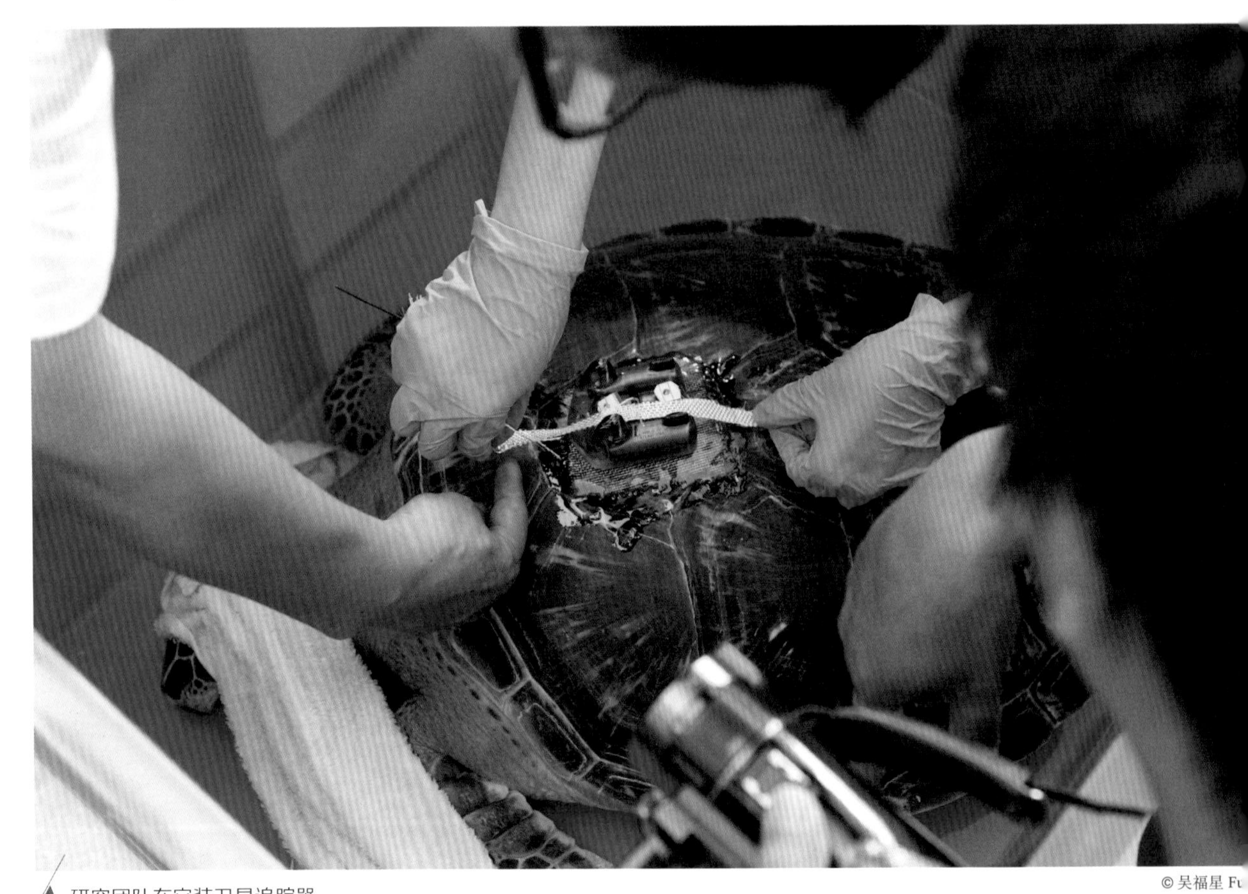

© 吴福星 Fu

▲ 研究团队在安装卫星追踪器

A research team fixed a satellite tag to the back of a green turtle

▲ 工作人员将幼龟放归海龟
Juvenile green turtle released

▲ 幼龟回到大海
Back in the sea

真实的故事 A true story

这只卫星定位追踪器编号 34474 的绿海龟是从非法交易者手中解救出来的。2018 年 5 月 23 日，它被成功放归大海。这只海龟从三亚海棠湾出发，先在海南岛附近兜兜转转。十几天后，它向西沙游去。最后，它进入北部湾海域并在越南海域消失。

This green turtle was seized from an illegal trader and successfully tagged (#34474) and released back into the ocean on 23rd May, 2018. The turtle was released in Haitang Bay, Sanya City, Hainan Province. It swam along the coastline of Hainan Island, migrated to the Xisha Islands and then entered Beibu Gulf. It eventually disappeared in Vietnamese waters after about two weeks.

▲ 编号 34474 海龟等待放归

The green turtle # 34474 ready to go home

▲ 装上卫星追踪器的编号 34474 的绿海龟
The green turtle with satellite tag number # 34474

工作人员采用正确的方式搬运海龟进行放归

Using proper methods to transfer turtle to the release site

© 北京市企业家环保基金会 (SEE Foundation)

发现受伤海龟后如何处理?

在岸上看到受伤的海龟，要第一时间联系当地公安或渔政部门，报告发现海龟的地点。在等待专业人员到场的过程中，可以用干净的湿毛巾轻轻地覆盖海龟的眼部（注意，千万不可遮蔽其鼻孔），这样可以帮助海龟减少应激反应。若有条件，尽量为海龟遮阳，并且不断用水保持海龟的背甲湿润，避免海龟被晒伤。搬动海龟时，可以握住海龟背甲靠近颈部和尾部的两端，但千万不要抓前后肢。这样既是为了防止被海龟打伤，也是为了避免伤到海龟。体型较大的海龟需要利用担架或其他辅助工具搬运。

What can we do if we find an injured marine turtle?

When you see an injured marine turtle on shore, contact the local public security or fisheries authority at once to report the location. While waiting for the authorities to arrive, you can cover the turtle's eyes with a clean, wet towel (do not cover the nostrils) to help reduce stress. If possible, provide shade for the turtle to prevent heat stress from the sun and keep the body moist with water. When moving turtles, grasp the carapace both in front and at the back (not the front or rear flippers). Larger marine turtles should be moved by a stretcher.

任何人都可以为保护海龟贡献力量

不干扰海龟上岸产卵（比如你晚上漫步沙滩时，突然看到海龟爬上岸，应该立刻蹲下保持安静，不要打扰海龟后续产卵）。不侵占海龟的产卵场。制止和举报偷窃龟卵的行为。

We all can contribute to the conservation of marine turtles.

Do not disturb nesting females on the beach. Do not occupy nesting beaches. Stop and report the illegal harvest of turtle eggs.

· 拒绝购买海龟，拒绝购买、使用、食用海龟制品。
· 看到非法交易海龟的行为，向公安报案。
· 减少使用塑料制品，不向海洋丢弃垃圾。
· 如果在野外发现海龟，不要打扰它们。
· 看到受伤的海龟，联系当地的渔政部门或者相关的救护机构。
· 将海龟的知识和保护海龟的重要性告诉身边的人。
· 人人行动起来，给海龟一个美好的未来。

· Refuse to buy marine turtles or buy, use or eat marine turtle products.
· Report to public security authorities if you discover illegal trade of marine turtles.
· Reduce the use of plastic products and do not throw trash into the sea.
· If you come across a marine turtle in the wild, do not disturb it.
· Contact your local relevant authorities or rescue organization if you see an injured marine turtle.
· Share your knowledge about marine turtles with those around you and the importance of protecting them.
· If everyone takes action, we all can make a beautiful future for marine turtles.

在海洋世界中，海洋动物们既是竞争者、合作者，也是制约者。海洋中的一切都以无比神奇又恰到好处的方式呈现，就像是一首和谐的协奏曲，演奏着大自然的不可思议。而人类的出现，却奏出了不和谐的音符。以海龟为例，海龟的濒危与人类有着直接关系。人类食用海龟卵和海龟肉，把海龟的龟甲和盾片当作药材使用，污染海洋环境……人类似乎忘记了，人和自然本就是一体的，人也只不过是地球生物圈中微不足道的一环而已。幸运的是，如今虽有人依然我行我素，但也有人开始为保护海龟、保护海洋积极行动起来了。

In the ocean, marine animals are competitors, collaborators and restrictors. The threats to marine turtles are directly associated with humans. We consume marine turtle eggs and meat, utilize their scutes as medicine and pollute the marine environment …… We forget that we are a part of nature.

第四章 Chapter 4

拯救海龟大作战 Let's Act to Save Marine Turtles

中国现存最大的海龟产床——南海西沙群岛

Protecting the largest nesting ground in China—the Xisha Islands

目前中国现存最大的绿海龟产卵场在隶属海南省三沙市的西沙群岛。由于远离大陆，人口密度低，开发程度低，它成了中国海龟产卵的一片净土。三沙建市以来，当地政府便开始加强对海龟产卵场的保护工作，并在 2015 年成立北岛海龟保护站，对海龟产卵进行持续的监测记录。七连屿从 2016 年有监测记录开始，每年均有超过 100 窝上岸产卵的绿海龟蛋坑，产卵岛屿主要包括北岛、南岛、西沙洲、南沙洲、中岛等。而西沙洲附近还分布有非常茂盛的海草床，这是成年绿海龟最喜欢的食物。据当地渔民说，在 4—6 月海龟产卵季到来之时，附近海域常常可以看到交配的绿海龟，西沙洲水域还经常能够见到觅食的绿海龟。晋卿岛和甘泉岛也均有绿海龟上岸产卵的记录，尤其是甘泉岛，近年来绿海龟上岸产卵的数量在逐年上升。目前三沙市海洋保护区管理局对整个西沙的海洋生态进行了统一的保护和管理，促进了海龟保护的系统化、科学化和社会化参与。

The largest nesting ground for green turtles in China is located on the Xisha Islands, Sansha City, Hainan Province. Since 2016, there have been several recorded green turtle nesting islands on Qilianyu cluster, with more than 100 nests recorded annually, including North Island, South Island, West Sand, South Sand and Middle Island. According to local fishermen, during the turtle breeding season (from April to June), green turtles can be seen mating in nearby waters. West Sand is surrounded by high seagrass meadows, the favorite food of green turtles. Other islands such as Jinqing Island and Ganquan Island also have high records of nests. The Sansha City Marine Protected Area Administration currently provides systematic management and protection measures to ensure a better future for marine turtles and their habitat.

鸟瞰七连屿

Bird's-eye view of Qilianyu cluster in Hainan Province

© 王静 Jing Wang

沙群岛的海草床是绿海龟的主要觅食场

eagrass meadows in Xisha Islands are major feeding
ounds for green turtles

© 宋稼豪 Jiahao Song

西沙北岛当地巡护员插的监测巡护牌，
第 101 号产卵标记（2017 年）

The nest label, No. 101 (2017), was inserted by
a local patroller at North Island, Xisha Islands

© 刘敏 Min Liu

中国海龟保护联盟：行动起来，保护地球的活化石

China Sea Turtle Protection Alliance (CSTCA): action for conservation of marine turtles

为了促进多方合作从而更好地保护海龟，由原农业部渔业渔政管理局指导，政府主管部门、多家公益组织、科研院校和爱心企业联合，于 2018 年 5 月 23 日世界海龟日共同发起成立了中国海龟保护联盟 (以下简称 CSTCA)。

In order to promote multiple parties to better protect marine turtles, CSTCA was founded on 23rd May, 2018, under the guidance of the Fisheries and Fishery Administration of the Ministry of Agriculture and Rural Affairs (MARA) of China, and joint government offices, NGOs, institutions and responsible enterprises.

▲ 2018 年 5 月 23 日，中国海龟保护联盟成立大会（海南三亚）
The CSTCA was founded on 23rd May, 2018 in Sanya, Hainan Province

CSTCA 的目标在于为海龟保护提供多方合作与交流平台以及科普宣教、救助放生、技术咨询等服务，促进全社会共同行动保护这些在地球上生存了亿万年的“活化石”。CSTCA 成立当日便放归了 5 只救护绿海龟。之后，通过联合北京市企业家环保基金会、永续全球环境研究所等不同的公益环保机构以及厦门大学、海南师范大学和华东师范大学等科研机构，还有相关的水族馆和救护站等，共同推动完成海龟的救护、暂养和放归工作。

The goal of the CSTCA is to provide a platform for cooperation and communication on marine turtle conservation, to provide services on public education, rescue and release and technical consultation for marine turtle conservation, and to promote conservation actions with all societies involved. On the founding day, 5 rescued green turtles were released. Since then, NGOs such as Global Environment Institute (GEI) and the Society of Entrepreneurs and Ecology Foundation (SEE Foundation), institutions such as Xiamen University, Hainan Normal University and East China Normal University, and public aquaria and rescue stations have worked together to complete rescue, holding and release activities.

© 范敏 Min Fan

▲ 永续全球环境研究所、北京市企业家环保基金会等公益组织配合农业农村部渔业渔政管理局开展海龟逃生装置（TED）测试，希望在渔民的捕捞作业中推广 TED 的使用，减少对海龟的误捕

GEI and SEE Foundation, together with other NGOs are cooperating with Fisheries and Fishery Administration to carry out TED tests, hoping to promote the use of TEDs in fishing operations and reduce bycatch of marine turtles

中国的公益组织在千年渔港海南琼海潭门镇，与日新村村民合作开展废弃渔具回收工作，减少主动和被动丢弃的废弃渔具成为海洋垃圾

NGOs in China are working with local fishing communities in Tanmen County, Hainan Province to carry out a fishing gear recycling project to reduce marine debris

真实的故事 A true story

2020 年 8 月 9 日上午，海南陵水县分界洲岛水域放归了 99 只救治康复的海龟。这 99 只海龟中，有 29 只是江苏司法机关在办案中移送救助、经专家救治和野化 训练后放生的，有 10 只身上被装上了卫星定位追踪器。这些海龟似乎也明白将要发生的事情，原本在水盆里还很安静的海龟到了甲板上后，仿佛闻到了熟悉的海水味道，开始舒展地扑腾着四肢，非常兴奋。在众人的关注下，一只只海龟依次从放归船甲板的坡道上顺势滑下，像利箭 一般，冲向久别的大海里。放生的海龟当日就开始发回卫星信号。

On 9th August, 2020, 99 rescued and recovered marine turtles were released in the waters of Lingshui County, Hainan Province. Among these, 10 turtles were fitted with satellite tags.

▲科研人员在海龟身上安装卫星定位追踪器
Researchers fixing a satellite tag on a green turtle

2020年8月9日，中国海龟保护联盟联合北京市企业家环保基金会（SEE基金会）等机构将99只救护的海龟放归大海

Together with the SEE Foundation and other organizations, the CSTCA returned 99 marine turtles to the ocean on 9th August, 2020

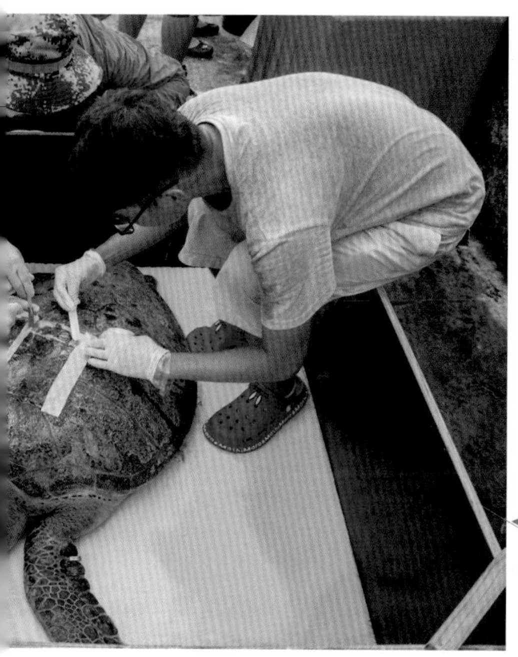

这只追踪器重127g，不足成龟一顿饭的重量，用来研究海龟的路径。科研人员正在用一种特殊胶水将其粘在海龟背上

The tag weighs only 127g, less than the weight of a meal for an adult marine turtle. It is used to study the marine turtle's migratory paths. Researchers are using special glue to stick the tag to the turtle's back

▲2021年7月28日，北京市企业家环保基金会（SEE基金会）联合中国海龟保护联盟送11只海龟回家
Together with the CSTCA, the SEE Foundation sent 11 marine turtles back to the ocean on 28th July, 2021

海龟对海水质量、温度、食物都有较高的要求，饲养不当很容易导致其生病、相互伤害甚至死亡。在海洋中，海龟以水母、鱼类、甲壳类、海草等为食，而人工饲养时则以鱿鱼等高脂肪食物为主，被贩卖收养的海龟中有的因此得了脂肪肝，甚至肝部发黑病变。而且，被人类捕捞驯养后，从自由深邃的大海到浅浅的鱼缸、水池，仅活动范围受限对海龟来说也是巨大的痛苦和伤害。由于人类的自私使得“长寿龟”无法长寿。

海龟野化训练是海龟救护、放归大海前重要的一环。放归大海前，海龟首先要在一个比较接近海洋的自然环境下野化一段时间。通过不投喂食物，让其自由觅食以获得基本营养需求，使海龟不同个体间形成生存竞争，恢复其求生本能，逐渐强化其在海洋中自我生存的能力。野化基地最好能毗邻出海口，潮汐落差明显，海水可以随潮涨潮落而进出围堰池，总体环境要接近海洋自然环境，且基地内要有较多的海藻、海草、水母、鱼、虾等海龟的主要觅食对象。最重要的是，野化基地外围的公共海域要一直有野生海龟活动的踪迹。最后，只有恢复了野外生存能力的海龟，才能放归大海。

Marine turtles have specific requirements for seawater quality, temperature and food, and improper care in captivity can easily cause illness, injury or even death. Marine turtles mainly feed on jellyfish, fish, crustaceans and seagrass in the natural environment. While in captivity marine turtles are fed high-fat food such as squid; this has resulted in some turtles getting fatty livers and even some showing hepatic melanosis (when the liver changes to a darker brown or black color). Moreover, marine turtles in captivity further suffer because they are kept in shallow and small tanks and pools. The selfishness of humans prevents "long-lived turtles" from living longer.

Wild survival skills training is an important part of evaluating the success of a marine turtle's rescue and release. Marine turtles should be taught to feed on their own, to restore their survival instincts and to gradually strengthen their ability to survive in the ocean. The rescue center should be as close to the sea as possible and provide various natural foods such as seaweed, seagrass, jellyfish, fish and shrimp. Most importantly, there should be indicators of wild marine turtle activity in the surrounding waters. In the end, only turtles which have regained their survival skills can be released to increase the survival rate in the wild.

法律法规为海龟保护建起铜墙铁壁

Laws and regulations set up protective walls for marine turtles

除了前面提到的《中华人民共和国野生动物保护法》《中华人民共和国刑法》，中国还颁布了《中华人民共和国水生野生动物保护实施条例》《水生野生动物利用特许办法》等一系列法规对包括海龟在内的珍稀濒危野生动物进行保护管理。中国各地也根据自己的具体情况出台了地方性的管理规定来保护海龟，例如广东的《广东省海龟资源保护办法》。

China has enacted a series of laws and regulations to protect and manage precious and endangered aquatic wildlife, including marine turtles. These include the *Wildlife Protection Law of the People's Republic of China*, the *Criminal Law of the People's Republic of China*, *Implementing Regulations on Aquatic Wildlife Protection of the People's Republic of China* and the *Licensed Measures for the Utilization of Aquatic Wildlife in the People's Republic of China* to name a few. Different provinces also released regional regulations to protect marine turtles under specific conditions, such as, the *Guangdong Province Marine Turtle Resource Protection Measures*.

《海龟保护行动计划（2019—2033 年）》提出海龟保护新挑战

Action plan for marine turtle conservation (2019—2033) provides guidance for marine turtle conservation

2019 年，农业农村部专门组织编制并发布了《海龟保护行动计划（2019—2033 年）》，确定了中国海龟保护的 7 项重点工作和 18 项重点行动。通过以下措施来联合各方的力量来共同推动海龟保护：1）建立健全保护体系与机制；2）加强海龟栖息地保护；3）稳步推进海龟收容救护，规范海龟人工繁殖与饲养；4）改善渔业作业方式，加强渔业监管；5）加强相关领域科学研究；6）加强保护宣传教育与公众参与；7）加强国际交流，建立国际合作机制。

In 2019, MARA released the China National Action Plan for Marine Turtle Conservation (2019–2033), with 7 key tasks and 18 key actions identified. To unite all parties to promote marine turtle conservation by 1) Building a comprehensive conservation network and system, 2) Enhancing habitat protection, 3) Improving marine turtle rehabilitation and formalizing artificial breeding & rearing of marine turtles, 4) Improving fisheries practice and reinforcing fisheries enforcement and monitoring, 5) Enhancing scientific research, 6) Promoting public education and engagement, and 7) Encouraging international exchange and collaboration.

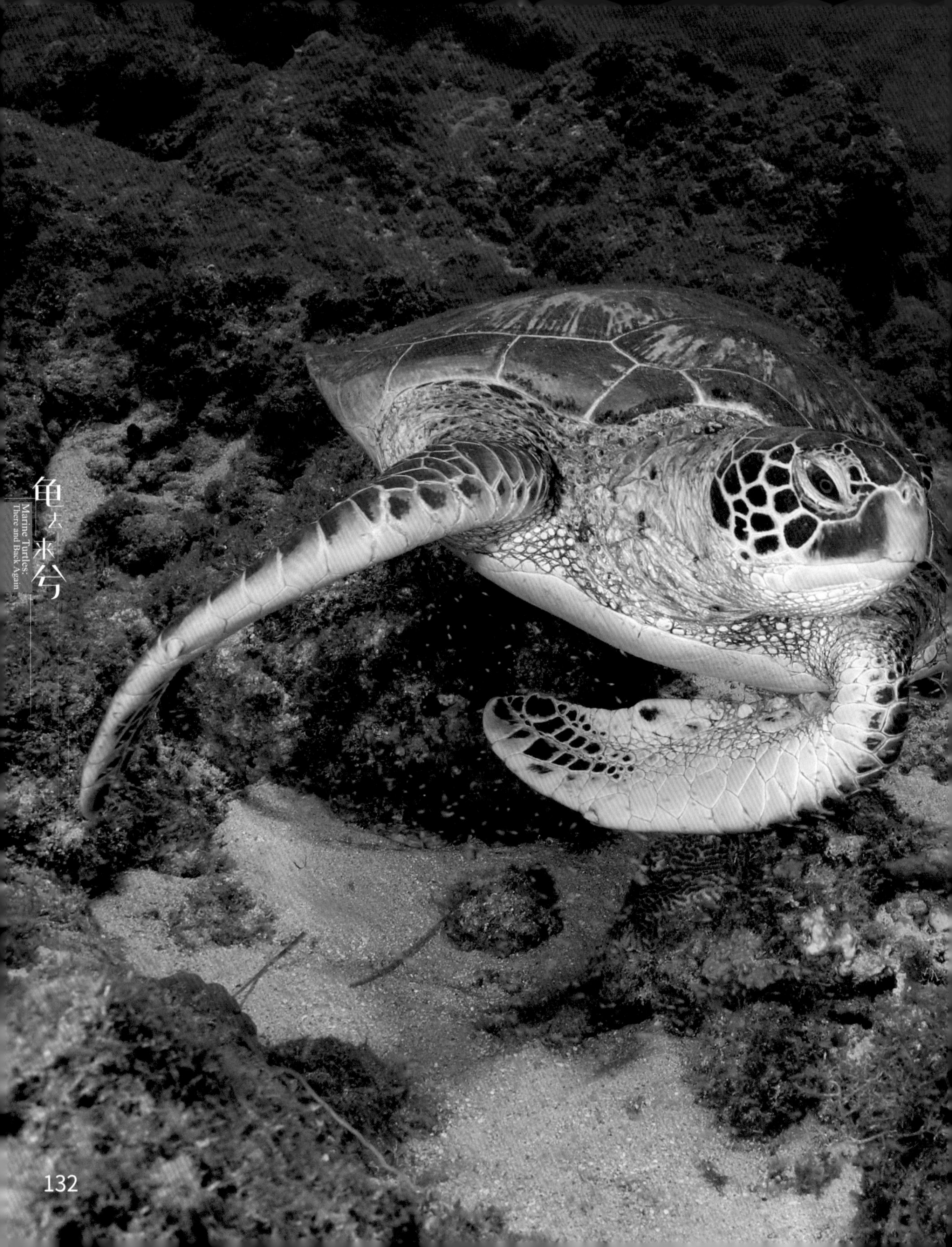

虽然无数物种已消逝在时光的边缘	Countless creatures have faded away over time
但海龟却以其特有的坚韧和顽强	Marine turtles
	With their unique resilience and hardiness
穿越历史的长河	Through the long process of history
从久远的过去来到我们身边	From the remote past time to meet us
看遍海枯石烂　历经桑田沧海	Significant changes have witnessed over time
	The sea goes dry and the rocks melt with the sun
它们是远古的信使	Marine turtles are the messengers of ancient times
把亿万斯年的悠扬史诗	The long epic of hundreds of millions of years
深深的镌刻在厚重的龟甲上面	Deeply engraved on the thick and heavy carapace
然而海龟	Yet, marine turtles
却正面临前所未有的苦难	are facing unprecedented hardship
非法捕捞　栖息地衰退　环境污染	Illegal fishing, habitat loss, environmental pollution
正是人类的傲慢和贪婪	All because of human arrogance and greed
为这些古老精灵的前路	Have set obstacles for the future of these ancient spirits
蒙上一层可叹又可悲的黯淡	Which are sadness and bleakness
万中有幸　为时不晚	It's not too late luckily
传承亿万年的故事	The story that has been passed through hundreds of millions of years
绝不该在我们手中谱下终篇	Shouldn't end in our hands
保护海龟正在行动	Protect marine turtles, we are taking actions
让你我携手并肩	Let's join hands
为海龟营造一个光明而广阔的明天	To create a brighter and broader future for marine turtles

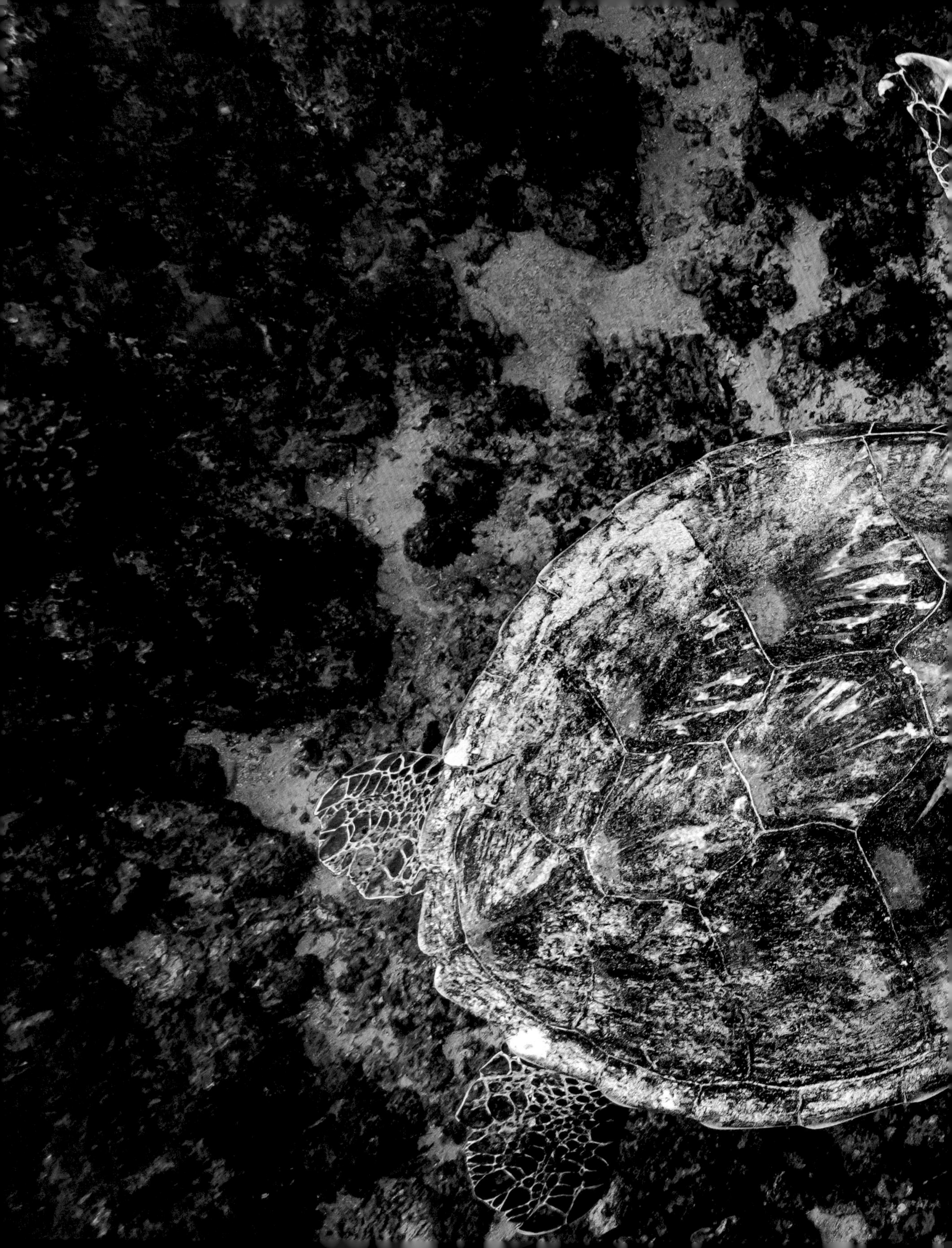

致 谢

ACKNOWLEDGEMENTS

撰写本书的过程对于我们来说也是一个系统梳理这些年关于海龟保护工作的过程。中国海龟保护所取得的成绩，是所有水野人辛勤努力的一个缩影。首先，我们要感谢所有渔政水野的工作者们在海龟保护、救护、执法过程中付出的辛勤努力；感谢农业农村部渔业渔政管理局，全国水生野生动物保护分会，海南省三沙市海洋保护区管理局，三沙市七连屿工委、管委会，三沙市永乐群岛工委、管委会，厦门大学，华东师范大学，海南师范大学等众多中国海龟保护联盟成员单位对于海龟保护工作的支持；感谢李彦亮、韩旭、江开勇、栗倩云、郭睿、姜波、刘立明、王春、邹志、黎明、陈芳、常英、李育培、陈燕华、史海涛、于洋飞、刘宁、林柳、杜宇、陈珉、张婷、张晶、张文婷、李晓菲、王昊、Steve Blake、杨袁筱月、贾语嫣、滕聿央和李沛聪等所有海龟保护工作者过去这些年对于海龟保护的倾情付出；感谢黄宏波、梁峰、王雄、陈山、许德壮、梁昌健、黄程等所有坚守在一线日常巡护的渔民朋友们。此外，我们还要感谢以下几家公益机构：在2010—2017年，保护国际基金会在海龟保护工作中给予的诸多前期支持工作；2018年到现在，永续全球环境研究所、北京市企业家环保基金会、野生救援（美国）北京代表处、问海自然保护中心等公益机构对于中国海龟保护联盟各项工作的支持。感谢廖慰南、黄文思、张冀强、金嘉满、张立、杨彪、季琳等公益机构的领导人对于海龟这项公益环保事业的大力支持。还有许多默默无闻的奉献者，正是因为有许多像他们这样辛勤付出的人士才有了这本书中提到的各项中国海龟保护的工作。

本书精美图片由盖广生、Ace Wu、张田、杨位迪、何云昊、王思宇、吴江、王静、王洁、梅云华、张晶、刘敏、张婷、吴福星、叶飞、范敏、刘孙谋、胡全君、李沛聪、曹志刚、宋稼豪、和杨子羿提供； 此外，一些新闻图片来自中国新闻网、央视新闻；一些网络图片来源于Pixabay.com和unsplash.com上的创作者们，还有一些照片来自北京市企业家环保基金会、永续全球环境研究所、三沙市海洋保护区管理局和问海自然保护中心。感谢他们为海龟留下的精彩瞬间，成就了本书的精美画面。最后，我们要特别感谢两位国际友人George H. Balazs和Yoshimasa Matsuzawa对于中国海龟保护工作的长期支持，尤其是Balazs先生已经在海龟保护领域工作超过50年，他丰富的经验，谆谆的教诲，默默无闻的付出，让众多中国的海龟环保人士受益良多，也帮助我们极大地推动了中国的海龟保护工作。感谢所有为海龟和海洋保护做出贡献的个人和机构。

Thanks to all the marine turtles' guards, because of their hard work, we have this beautiful book. Special thanks to George H. Balazs and Yoshimasa Matsuzawa for their generous help with China marine turtle conservation. Thanks to Guangsheng Gai, Ace Wu, Tian Zhang, Weidi Yang, Yunhao He, Siyu Wang, Jiang Wu, Jing Wang, Jie Wang, Yunhua Mei, Jing Zhang, Min Liu, Ting Zhang, Fuxing Wu, Fei Ye, Min Fan, Sunmou Liu, Quanjun Hu, Peicong Li, Zhigang Cao, Jiahao Song and Ziyi Yang, and the website of www.chinanews.com, news.cctv.com, pixabay.com, unsplash.com for providing the beautiful photos seen throughout. Thank you to all the individuals and organizations working for marine turtle conservation and marine conservation!

《龟去来兮》编委会
Editorial Board

2021年9月
September, 2021